Mapping Work Processes

Also available from ASQC Quality Press

LearnerFirst™ Process Management software
with Tennessee Associates International

Process Reengineering: The Key to Achieving Breakthrough Success
Lon Roberts

Reengineering the Organization: A Step-by-Step Approach to Corporate Revitalization
Jeffrey N. Lowenthal

Principles and Practices of TQM
Thomas J. Cartin

The ASQC Total Quality Management Series

TQM: Leadership for the Quality Transformation
Richard S. Johnson

TQM: Management Processes for Quality Operations
Richard S. Johnson

TQM: The Mechanics of Quality Processes
Richard S. Johnson and Lawrence E. Kazense

TQM: Quality Training Practices
Richard S. Johnson

To request a complimentary catalog of publications, call 800-248-1946.

Mapping Work Processes

DIANNE GALLOWAY

ASQC Quality Press
Milwaukee, Wisconsin

Mapping Work Processes
Dianne Galloway

Library of Congress Cataloging-in-Publication Data

Galloway, Dianne
Mapping work processes/Dianne Galloway.
p. cm.
Includes bibliographical references and index.
ISBN 0-87389-266-6 (alk. paper)
1. Work design. 2. Flow charts. I. Title.
T60.8.G35 1994
658.5'42—dc20 94-13349
CIP

10 9 8 7 6 5 4

ISBN 0-87389-266-6

Acquisitions Editor: Susan Westergard
Project Editor: Kelley Cardinal
Production Editor: Annette Wall
Marketing Administrator: Mark Olson
Set in Franklin Gothic and Optima by Linda J. Shepherd
Cover design by Montgomery Media, Inc.
Printed and bound by BookCrafters, Inc.

ASQC Mission: To facilitate continuous improvement and increase customer satisfaction by identifying, communicating, and promoting the use of quality principles, concepts, and technologies; and thereby be recognized throughout the world as the leading authority on, and champion for, quality.

For a free copy of the ASQC Quality Press Publications Catalog, including ASQC membership information, call 800-248-1946.

Printed in the United States of America

 Printed on acid-free recycled paper

Contents

Preface

More than a decade ago—when my quality library occupied just a few inches of bookshelf—my colleagues and I knew that a key to translating quality values, philosophy, and principles into measurable results required, in part, a broad and profound understanding by employees of the work within their organizations. In hundreds of classroom hours we tried dozens of techniques to help groups of people document what they knew about their jobs—to commit the details of work sequences to paper.

Mapping, a methodology

Flowcharting, with its large-scale visual format, seemed to hold the most promise. So over the years, a step-by-step method has evolved—a method for getting what's in people's heads onto paper in a way that 1) can be quickly learned, 2) is appealing and energizing, and 3) results in a usable product. Early on I dubbed this method *mapping*. Though traditional flowcharting is its inspiration, mapping abandons some of the tradition to better serve the goal of simplicity and directness.

Mapping is merely an enabler—a means to a more important end. It is a vehicle for expressing and releasing the knowledge, creativity, and energy that lies within every group, regardless of its position or level within an organization. And while the mapping activity is valuable by itself, the second challenge was (and continues to be) to compile and validate specific ways to *use* the visual map to inspire meaningful, creative change. There are a baker's dozen improvement ideas listed and summarized in the last two chapters. They are extracted from our longer list of 25 improvement techniques, but these 13 have as their prerequisite a completed process map.

While other excellent books in the area show detailed flowcharts of business processes, this book itemizes the *process* of mapping—the how-to, step-by-step. Its purpose is to help groups avoid some of the procedural errors that typically occur when they invent their own ways of flowcharting. For example, most groups inevitably step into the "should be" before analyzing the "what is." Individuals within a group often discover that they do parts of the job differently from one another. How then, can the group accommodate and display these differences? The easy solution is to make premature decisions about what the one best way *should* be. The method presented here encourages groups to separate the "what is" from the "should be" so that developing improvements gains the careful, focused consideration it deserves.

Examples and illustrations

A second aspect of this book needing some explanation is my selection of examples. Three examples (setting a table, getting gas for your car, and getting ready for work) illustrate the methodology. I am always asked, "Why not use generic, business-related examples, such as processing paperwork, a simple assembly process, or a customer service example? Wouldn't readers relate to business examples better? Why insult their intelligence with vacuous examples?"

The answer is grounded in learning theory. If acquiring *knowledge* (facts and theory) is the objective, the business-related example works best because its content is the source for learning. But when learning a skill—how to do something—the reader's focus must be directed toward what is being done with the example (the actions and operations), and away from its subject matter. Using business examples to develop skill invariably misdirects a learner's attention. We know this from long, sometimes painful, experience. A skill-building example must be:

- Familiar to everyone; no study should be required to understand it well
- Sufficiently complex to offer a substantial platform for application
- Inconsequential; it should not generate arguments from authorities or experts who can assert superior knowledge about its content

Unfortunately, even the broadest business examples are too specialized to meet these requisites. But examples from everyday life can. The content of the examples is purposely irrelevant but familiar. Ignore the content and focus on the process of mapping and the issues that arise from its application to specific work processes. That's what's really important.

Intended audience, users

Further, use of common examples permits learning of the method by a much wider audience. The methodology has been successful in a range of organizations from service (banking, health care, utilities) to manufacturing (engineering, assembly, maintenance), to public sector (education, government groups) at all levels of education.

While most organizations recognize the need for workforce training if they are to meet the quality challenge, they are also searching for more cost-effective alternatives to facilitator-led, classroom delivery. These materials will provide, I believe, one such alternative.

My thanks to the scores of classroom participants who submitted to my constant tinkering—who cheerfully (usually) tried new materials, challenged the taxonomy and examples, and invented better ways of seeing. Because of them, I am able to document the mapping methodology in a do-it-yourself format for teams committed to improving the quality and efficiency of their own work processes.

Dianne Galloway
January 1994

1

Introduction to Mapping

What's a map?

The map of a work process is a picture of how people do their work. A town map (Figure 1.1) shows many possible paths from a chosen starting point to the desired destination and may show various features—such as shop or school locations—along the way.

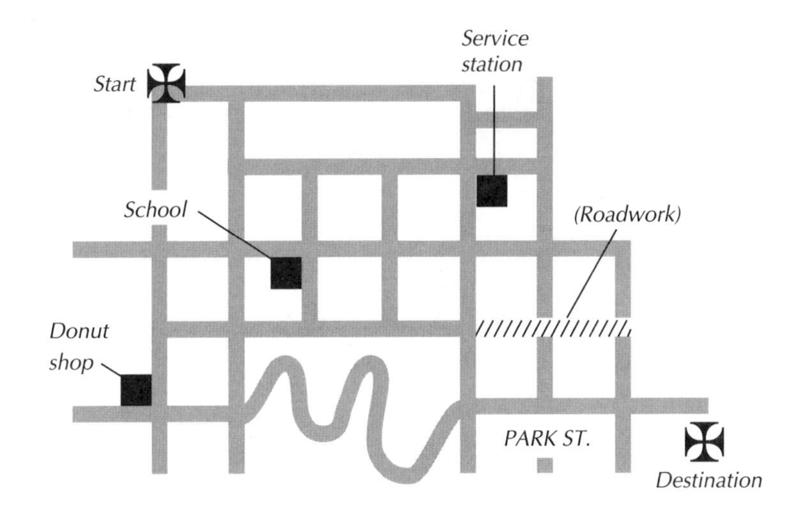

Figure 1.1. Alternative paths.

PROCESS MAP*
A graphic representation of a process, showing the sequence of tasks; uses a modified version of standard flowcharting symbols.

Different drivers will plot their cross-town journey over different routes based on their varying needs to fill the gas tank, buy a donut, avoid roadwork, or pick up the kids at school. Work processes are similar to road networks in that different people choose different routes to the same destination for different reasons. This book will show you, step by step, how to draw an accurate map of your work process showing alternative paths and methods. A good map is the foundation for continuous quality improvement efforts in which you analyze and agree on the most efficient routes to take under various circumstances. The technique we'll use is a modified flowchart.

**Key definitions appear in the margins.*

Although this book is intended for self-directed groups of people who share a work process, it's possible (but more difficult) for an individual to complete a good process map. Why more difficult? Because mapping is an exercise in looking for alternatives and—later—judging which are best under certain conditions. Individuals are more likely to chart their own familiar path and declare it "best" without considering all the different circumstances and alternatives.

Organization of book

Chapters 2 through 6

The next five chapters are devoted to showing you how to create a map of your process as it now exists. Within each of these chapters, you'll find:

- *A brief segment of information about the topic, new terms, and/or illustrations of flowcharting symbols and conventions.* Begin each chapter by reading the information segment. Either read it aloud together or assign the reading to be done before the meeting begins.

ANALYZE

- *Sample process maps in progressive stages of completion.* These examples show you what you're aiming for with your own process at each step along the way. There are some questions for you to work through—just to make sure you've understood the key concepts before starting to work. Answers to the questions follow. We recommend that you work through the questions together, as a group.

- *Step-by-step instructions on how to proceed.* Sometimes more information will be included to help you complete the instruction. Follow these instructions fairly closely—they're based on our experience helping dozens of teams map their work processes.

MAPPING
The activity of creating a detailed flowchart of a work process showing its inputs, tasks, and activities, in sequence.

❮ Because quality improvement has its own specific language made up of familiar words used somewhat differently, you'll find key definitions highlighted in the margins for quick reference. There is also a glossary of these terms at the end of the book on page 83.

Chapters 7 and 8

These two chapters are devoted to listing and describing a number of techniques showing how to use the map once you've created it. Thus you'll use the map to improve your process.

Chapter 7 discusses five techniques you should apply to *every* process, without fail. Chapter 8 lists and describes eight other techniques that are optional but recommended.

Materials

Figure 1.2 is a very simple map (or agenda) listing all the steps for mapping a work process (it is keyed to the chapters and activities in the book). As you begin your first session, check the agenda for the materials you'll need for the section you'll be working on. Then, find a large, flat work space—like a wall, white board, or table. You'll cover the space with flip chart paper onto which you'll attach stick-on notes or index cards. You'll also need an assortment of markers, pencils, and erasers. (Yes, you're going to make mistakes—no doubt about it.)

Time requirements

If it's your first time mapping it'll take you at least a day—perhaps two—to create the map, depending on your tolerance for detail. Applying and using the improvement techniques will take anywhere from a week to a decade or more (we're talking continuous improvement, remember).

Mapping is demanding work. Where possible, we recommend that teams work in half-day sessions—about the right length to sustain a group's attention. Shorter, two-hour sessions work well for some groups, but others swear by the intensive, two-day workshop format.

Your objective

The map you create will represent the process as it is now, with all its flaws and inefficiencies. It will be a working document—a means for getting to other, valuable improvement activities. Therefore, don't have as your chief objective a gorgeous document that will impress a lot of people—a document you won't want to change, mark up, and revise often.

Last, many people have said that, while the final map is a great tool, the real value of the exercise is in its creation. The discussions required to create the map help team relationships and increase the participants' understanding of how other people do things. So don't rush. Keep your eye on the final output, but don't shortchange what the exercise itself can do for you.

Meeting roles

Effective, efficient meetings are those that involve all participants and achieve the tasks or objectives in a reasonable amount of time. To ensure good results, we recommend that three roles be assigned for each meeting.

Facilitator

The facilitator leads the meeting. He or she ensures that discussions stay on track, that all members participate, and that the activities move along at a pace that is comfortable for the group. The facilitator role can be assumed by anyone in the group.

Scribe

The scribe records ideas on a flip chart or white board and makes sure that ideas are not lost. The flip chart is usually the focal point as the group works through various tasks. We don't recommend combining the roles of facilitator and scribe because it slows the group's pace.

Timekeeper

The timekeeper helps the team estimate the time required for each agenda item and monitors the group's progress against the estimate. From time to time, he or she may announce that there are just 10 (or 20) minutes remaining. If the group requires more time, the timekeeper helps it set a new target. The timekeeper should record target and actual times on a posted flip chart so that teams can learn to estimate more accurately as they proceed.

Getting organized

Use the "Team Roster," to record names and telephone numbers of team members along with meeting dates and attendance information. We've also included a few points about selecting the right people for the team.

■

Team Roster

Name	Telephone	Meeting Dates/Attendance					
1							
2							
3							
4							
5							
6							
7							
8							
9							
10							
11							
12							

Questions and answers about team selection:

- How large should a work team be?

The ideal size is five to eight participants. Fewer than five limits the range of ideas. More than eight leaves some people standing around, unable to participate fully.

- Who should be on the team?

The people who do the work—the people who know the detail of the job and how it is done. While managers may participate, they often don't have enough knowledge (or desire) to get down into the murky details of the day-to-day work. Of all the recommendations on this page, this is the most important, and ignoring it is the number one reason for mediocre results.

- What if there are a lot of people who do the same work? Which of them should we select for the team?

Ask for volunteers. That makes it clear to everyone what the selection criteria are and avoids speculation about the politics of selection. Also, volunteers are less likely to grumble about the time away from their regular duties because people tend to find the time to do the things they want to.

- What level of education is required to do a good job mapping?

None that we've found. It helps to be able to read English of course, but we've watched groups without high school education, groups of English-as-a-second-language speakers, and groups of Ph.D. scientists. Adequately motivated and supported, they all do a fine job—different, but fine.

- How about including suppliers and customers on the team?

Great idea! It's probably best, though, to wait until you have a good start on the map, then ask them to react to it and help you finish it off. The early part of the mapping process is likely to be excruciating for those not closely involved in the work steps.

- Will people really do all this stuff?

Absolutely. People just love talking endlessly about how they do their jobs and hearing how others operate. So long as team members believe that what they're doing matters and that they're not under pressure to produce too fast, few teams bail out before finishing.

- Don't we need an "expert" to facilitate?

No. The real experts are the workers. This book will provide enough structure to allow groups to work quite independently. Part of empowerment is having faith in people's ability to use their heads. Introduce an expert, and team members will believe it's the expert who has the best answers. Not so.

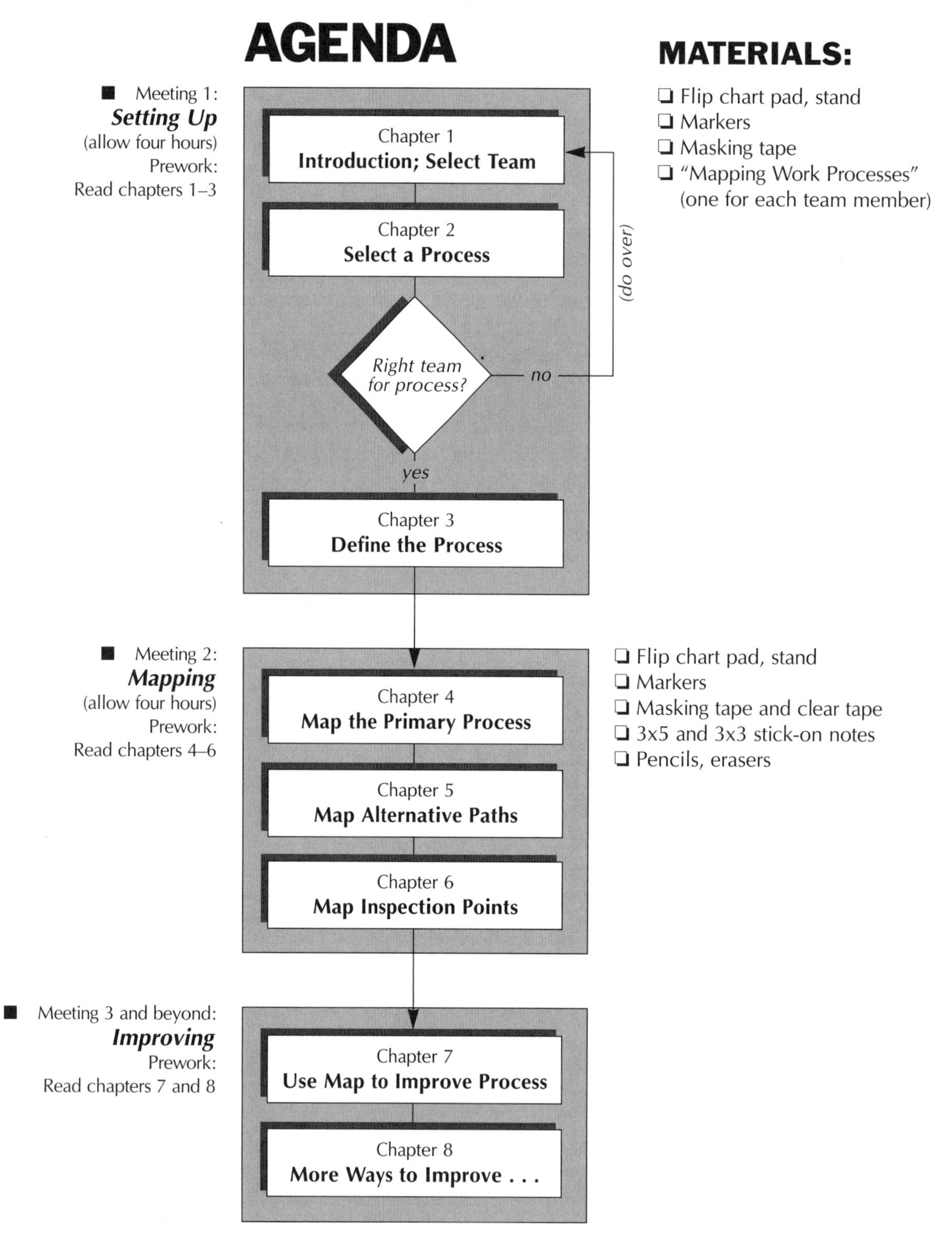

Figure 1.2. Agenda of steps for mapping a work process.

2

Select a Process

Your first agenda item is to select a process to map and—ultimately—to improve. But even if you've been handed a process to work on by someone else, you'll need to know a little about the nature of work processes and how they may differ from one another.

What's a work process?

A work process is made up of *steps, tasks,* or *activities* (we'll use these three terms interchangeably) and has a beginning and an end. Using inputs, it produces either a tangible product or an intangible service as its output. The process adds value to the inputs. At its simplest:

INPUT → PROCESS → OUTPUT

Applying the model, our drive across town might look something like Figure 2.1.

PROCESS
A sequence of steps, tasks, or activities that converts inputs to an output. A work process adds value to the inputs by changing them or using them to produce something new.

INPUT
The materials, equipment, information, people, money, or environmental conditions needed to carry out the process.

OUTPUT
The product or service that is created by the process; that which is handed off to the customer.

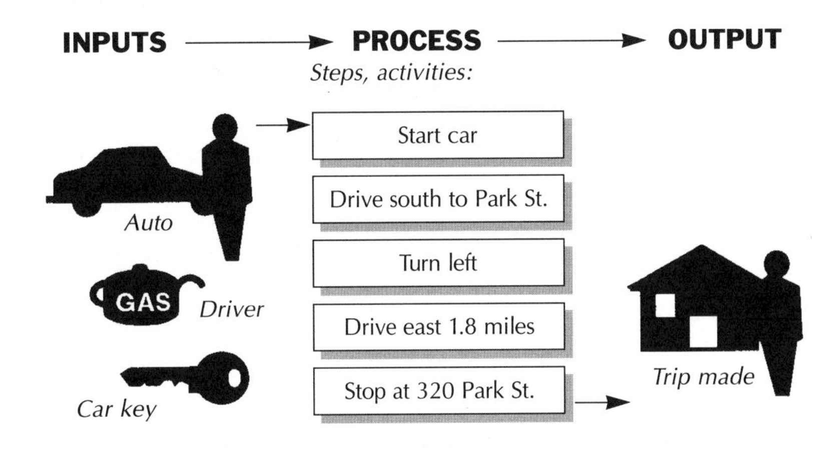

Figure 2.1. Inputs to outputs.

You can probably think of other important inputs to this process. Likewise, the number of steps could be far greater. The output of this process is a service rather than a tangible product. Tangible products are those that occupy space—they have height, weight, color, smell—and you can ship them off to a customer. Services are things you do for someone, such as drive them somewhere, solve a problem for them, or fix something for them.

Figure 2.2 shows another familiar example of a process with its inputs and (tangible) output:

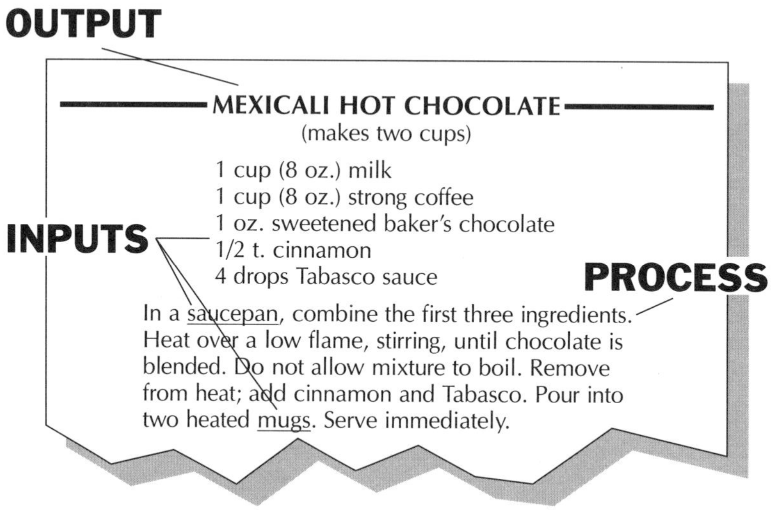

Figure 2.2. Recipe as a process.

Sample processes

The following is a laundry list of typical processes found within organizations. Browse through the list and check those processes that are found in your organization. Typically there are hundreds of different processes within a single organization.

For external customers

- ❑ Selling a product or service
- ❑ Repairing or maintaining a product
- ❑ Processing warranties
- ❑ Delivering or distributing products
- ❑ Billing
- ❑ Answering customer inquiries
- ❑ Manufacturing
- ❑ Entering orders
- ❑ Managing projects
- ❑ Preparing annual report

Internal, support processes

- ❑ Filing patents
- ❑ Conducting basic research
- ❑ Cleaning
- ❑ Maintaining grounds, facilities
- ❑ Conducting training classes
- ❑ Distributing the mail
- ❑ Answering telephones

Management processes

- ❑ Budgeting
- ❑ Approving travel
- ❑ Coaching/appraising
- ❑ Developing peoples' skills
- ❑ Setting objectives
- ❑ Communicating
- ❑ Hiring/firing
- ❑ Obtaining resources: people, money, materials, equipment
- ❑ Creating reports and memos

EXTERNAL CUSTOMER
User of an organization's overall product or service who is not a member of the organization.

INTERNAL CUSTOMER
User of products or services who is a member of the organization.

Some processes deliver their outputs to external customers, while others produce for customers who are other employees in the organization. Probably you checked more processes toward the end of the preceding list than at the beginning. This is because most organizations have similar processes internally, but differ in processes to external customers—which is what distinguishes one organization from another.

How many processes?

Most people can identify about three to eight different processes that they're a part of. Some managers will find that they work with a dozen or more processes. More than likely, you can identify two or three that occupy most of your time or that are the most important. Where you have a choice, you may want to select a process that serves your organization's external customer—because that's the source of customer satisfaction.

From whose point of view?

One of the first puzzles presented to groups of employees is how their level in the organization affects what they believe their processes to be. For example, Figure 2.3 represents a large auto repair shop. If we ask the owner (or top executive) to list processes, the list might look like the functional (departmental) organization of the business.

MACRO PROCESS
Broad, far-ranging process that often crosses functional boundaries (for example, the communications process or the accounting process). Several to many members of the organization are required to accomplish the process.

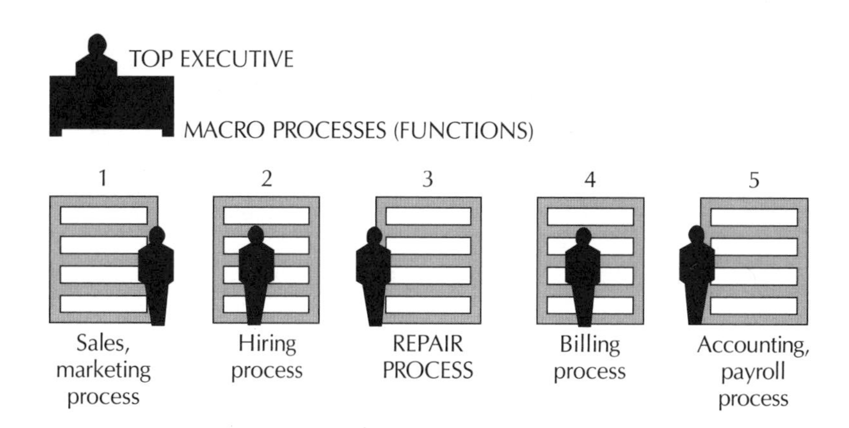

Figure 2.3. Functional view from the top.

But if you ask the manager or supervisor in the repair shop what the business processes are, he or she might look at it as in Figure 2.4—the repair function made up of a number of different, smaller processes.

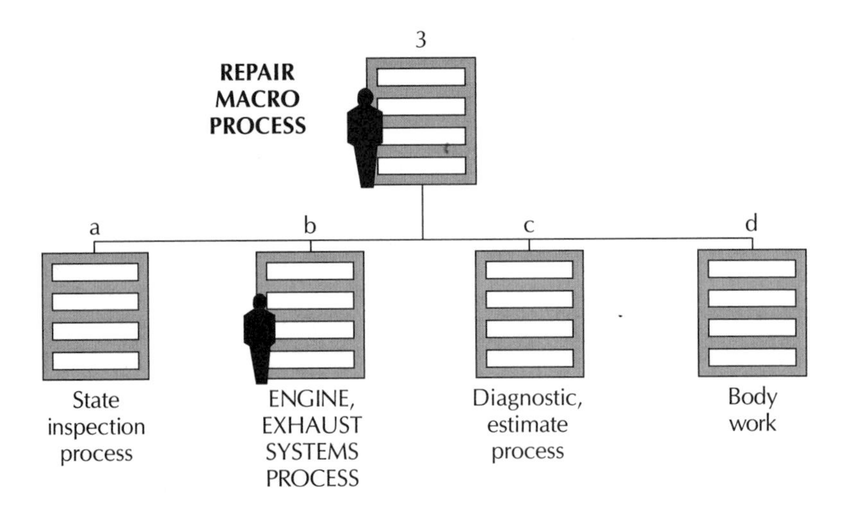

Figure 2.4. View from the middle.

Ask a senior mechanic about processes, and he or she will list even smaller blocks of work as shown in Figure 2.5—what we call *micro processes.*

MICRO PROCESS
A narrow process made up of detailed steps and activities. Could be accomplished by a single person.

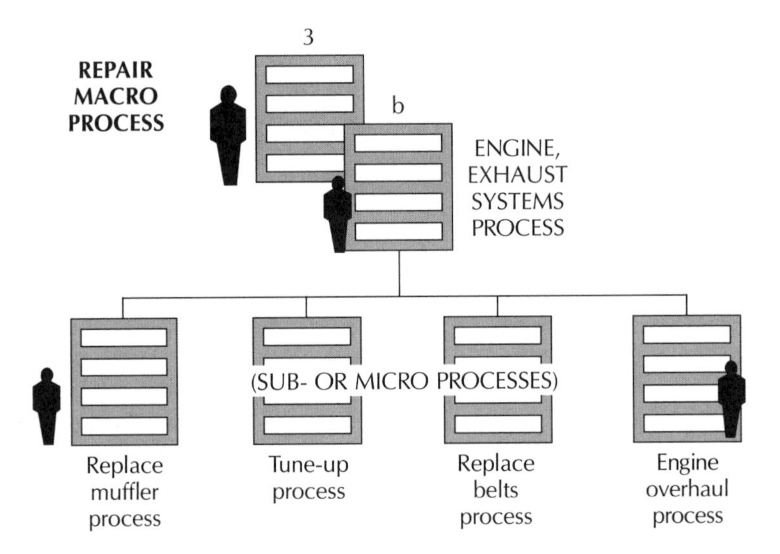

Figure 2.5. View from the micro level.

In other words, there are macro (big) processes that can be broken into smaller and smaller micro processes as various levels of specialization are accounted for. Thus a process is defined by the person whose process it is—the person who creates the output. So long as a group or individual has responsibility for the process—from start to finish—it can be considered a self-contained process regardless of its magnitude.

An important distinction because . . .

When executives embark upon process improvement, it's not surprising that they tend to identify large, cross-functional processes for examination; that's how they see things. Without a commitment to reengineer the entire business, they can find the results of such efforts disappointing because of the size and complexity of the task. An alternative is to begin with smaller, micro processes which, when improved one by one (by the level of people who know the details best), add up to significant cross-functional macro improvements. Perceiving the difference between macro and micro processes allows you to break down broad ideas for improvement into their smaller, more manageable parts.

As you might imagine, every organization is comprised of hundreds—even thousands—of interlocking processes. And this is the power of process improvement. To the extent that every employee is empowered to improve his or her process, the entire business can be optimized, bit by bit.

The customer's process

Sometimes it's helpful to map a process from a customer's point of view rather than that of the employee. An example: a team at a large hospital selected the admissions process for improvement. Instead of mapping the tasks performed by the hospital personnel to admit a patient, they mapped the routes, tasks, signatures, and other requirements the *patient* needed to perform. The team was horrified to find more than 40 steps and a half-mile walk was required of each patient!

Inventing new processes

Finally, we're often asked if mapping methodology can be used to plan and create *new* processes. Absolutely. Everything applies and contributes to a thorough planning effort.

In summary

- Work processes convert inputs to outputs. They add value to the inputs.
- Some outputs are delivered to external customers, others to internal customers. Many of the processes that produce outputs to internal customers are the same, organization to organization.
- Large, cross-functional processes that involve lots of people may be called *macro processes.* Executives and top managers tend to see and identify macros for improvement.
- Smaller, local processes can be called *micro processes.* Line employees tend to see the work as micro processes.
- Every organization has hundreds—even thousands—of work processes.
- One way to tackle large macros is to break them into their component micro processes and improve the micros, one by one.

Now, follow these steps to select and/or judge the process you will work on.

❶ Using the following Process Selection Matrix (Figure 2.6), enter the key business objectives of your company or department (depending on what's available) down the left side of the matrix. Typically, business objectives have to do with increased customer satisfaction, increased market share, a financial objective, and perhaps an employee satisfaction objective. Others may include safety, prestige, growth, and others.

❷ On a flip chart, brainstorm a list of processes for which you have responsibility. After you've finished brainstorming, make sure each fits the definition of a process. Adjust, revise, and reword each idea until you have between four and 10 processes.

❸ Enter your list of processes across the top of the matrix.

❹ Rate each process against each business objective by assigning each a value from 5 (process has a very high impact on business objective) to 1 (process has little impact on business objective). Work by rows, horizontally. Sum each column.

CONSENSUS
Agreement, harmony, compromise. A group decision that all members agree to support, even though it may not totally reflect individual preferences. Consensus is possible when diverse points of view have been heard and examined thoroughly and openly.

❺ Two or three of the processes will pop out as having the greatest impact on the business. Using any of the consensus techniques, select one of the processes to map. To help you make this final selection, consider the following list of criteria.

Key Business Objectives ▼ / Processes ▶	a.	b.	c.	d.	e.	f.	g.	h.
1.								
2.								
3.								
4.								
5.								
Sum ▶								

Figure 2.6. Process selection matrix.

Criteria for judging your selection

Once you've narrowed your processes down to just a few, here's a short checklist of criteria for selecting a process that is most likely to lead to a successful outcome. If you must answer "no" to two or more criteria, you should consider selecting another process that gets a better score.

yes no

☐ ☐ *The process fits the definition of a process;* it has an output, a customer, a beginning, and an end.

yes no

☐ ☐ *The process is small/simple enough in scope to be appropriate for a first project.* For example, if you select "world hunger" as your first project, you can get hopelessly bogged down and abandon your efforts. The "customer satisfaction process" is a "world hunger" kind of topic because it reaches into every department, every office, and every desk in the organization. More appropriate would be "customer complaint resolution," "measuring customer satisfaction," or "employee training in customer satisfaction."

yes no

☐ ☐ *The output or process has an impact on external customers.* Ask yourselves, "If this process were abolished, would it have any effect on the organization's customers?" Process improvements that create a better place for you to work should have lower priority—after external and internal customers are attended to.

yes no

☐ ☐ *Managers and executives will be sufficiently interested in the results of your work to give support.* Management usually cares most about budgets, safety, and customer satisfaction. Few of them lose much sleep over topics such as long lines in the cafeteria or the clarity of photographs in the newsletter.

yes no

☐ ☐ *The process is something this group knows about and has the authority to change.* If you find yourselves talking about what other people should do (". . . if only they'd change this or that . . ."), you're targeting someone else's process. You can't improve other people's processes for them. They'll tell you—with some justification—to go jump in the lake. Even if you come up with sensible, elegant improvements, you'll have trouble implementing them.

yes no

☐ ☐ *The process is not a solution to some problem.* Someone (such as your manager) may have requested that you implement a ready-made solution. While fixing the problem may be important, the activities in this book will confuse rather than help you.

■

3

Define the Process

Your next agenda item is to define your process. In a broad sense, this means understanding where your process fits into the larger organization/division context. Specifically, it means naming your customer and the output of your process—two concepts you're already familiar with.

In addition, you'll learn some new terms and definitions that will help you sort out important links between your work group and other people and processes within your organization: *process owners, process participants, stakeholders, process boundaries.*

❶ Complete the blocks on the following pages. Use the accompanying definitions to help you arrive at answers. To judge each of your conclusions, ask "Does our response conform to the definition in every way?"

❷ You may work on the elements *in any order.* We suggest you try them in the order presented, but feel free to skip around.

❸ Work on each element until the group is able to reach consensus.

❹ *Do not rush to decisions.* Part of the value of this item is the thoughtful discussion it can generate—leading to a deeper understanding of the impact your process has on the rest of the organization.

❺ Record your work; post it in a visible spot.

■

◆ State the OUTPUT of the process

✎

OUTPUT
The product or service that is created by the process; that which is handed off to the customer.

Outputs should be expressed in a noun/verb format—for example, "machines serviced," "orders logged," "reports submitted." Other descriptors may be added that clarify and limit, such as "quarterly outlook report submitted." The reason for expressing the output in noun/verb format is that it forces you to consider both that which is produced and the action you take.

....
noun

....
verb

◆ List the CUSTOMER(S) for your output

✎

CUSTOMER(S)
The person or persons who USE your output—the next in line. Whether your customers are internal or external, they use your output as an input to their work process(es).

List them by name where possible. In some cases, it's helpful to identify the chain of customers that receives your output.

....

....

....

....

◆ List your customer's REQUIREMENTS of your output

✎

REQUIREMENT(S)
What your customer needs, wants, and expects of your output. Customers generally express requirements around the characteristics of timeliness, quantity, fitness for use, ease of use, and perceptions of value.

....

....

....

....

◆ List the PROCESS PARTICIPANTS

✎

PROCESS PARTICIPANTS
The people who actually do the steps of the process—as opposed to someone who is responsible *for the process, such as the process owner/manager. For example, if you use subcontractors to produce the product, and you don't do the work yourself, the subcontractor is the process participant and you are the owner/manager.*

Process participants can be listed by name or by job title, provided all employees with the job title perform the process.

....

....

....

....

....

◆ List the PROCESS OWNER

PROCESS OWNER

The person who is responsible for the process and its output. The owner is the key decision maker and can allot organization resources to the process participants. He or she speaks for the process in the organization. That is, if someone says, "How come those California people aren't selling enough equipment?" the process owner—probably a District Sales Manager on the West Coast—would have to come forward to answer.

✎

....

....

◆ List the STAKEHOLDERS

STAKEHOLDER

A process stakeholder is someone who is not a supplier, customer, or process owner, but who has an interest in the process and stands to gain or lose based on the results of the process. Most processes have a number of stakeholders—such as senior managers from other departments or even government agencies.

List stakeholders either by name or by function, or both.

✎

....

....

....

....

....

....

◆ Agree on the PROCESS BOUNDARIES

PROCESS BOUNDARIES

The first and last steps of the process. Ask yourself, "What's the first thing I/we do to start this process? What's the last step?" The last step may be delivery of the output to the customer.

Note that you may come back and change the boundaries later, based on your flowcharting work.

✎ First step (an action)

✎ Last step (an action)

◆ INPUTS and their SUPPLIERS

INPUT

The materials, equipment, information, people, money, or environmental conditions that are required to carry out the process.

SUPPLIER

The people (functions or organizations) who supply the process with its inputs.

You'll be listing suppliers and their inputs later, but here are the definitions.

4

Map the Primary Process

PRIMARY PROCESS
The basic steps or activities that will produce the output—the essentials, without the "nice-to-haves." Everyone does these steps—no argument.

The *primary process* is the backbone of your complete work process. It consists of the essential steps or activities that must occur to produce your output. From the following activities, you'll begin a flowchart of your primary process.

Steps, activities

Each step, task, or activity within a flowchart is depicted as a rectangle. Figure 4.1 shows three steps of the "cleaning up after dinner" process.

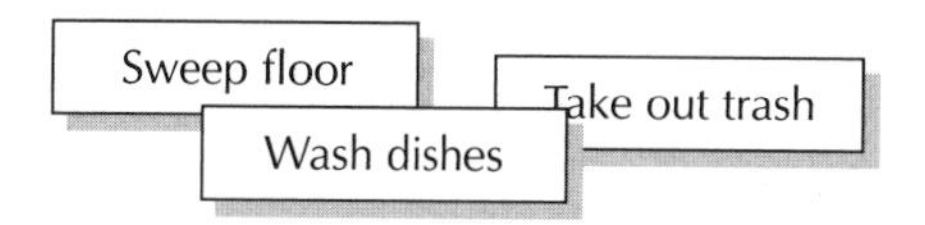

Figure 4.1. Various after-dinner steps.

Then, as shown in Figure 4.2, activity rectangles are placed in the *sequence* in which they occur.

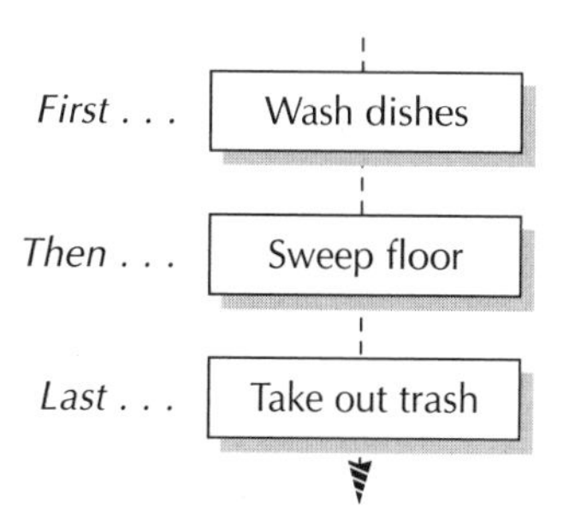

Figure 4.2. Steps in sequence.

Arrows showing the direction (sequence) of the tasks are dotted to represent lightly penciled, temporary lines. We'll want to move things around and erase before we're finished. Inked lines are less flexible.

Inputs, shown in Figure 4.3, are drawn as parallelograms linked to the step where they are used. Likewise, the output appears in a parallelogram.

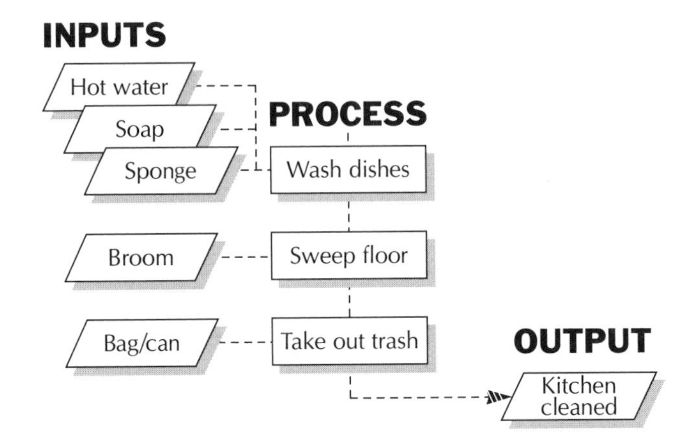

Figure 4.3. Inputs and outputs.

Sometimes the tasks in a primary process are shared by two or more people, creating a *parallel process,* shown in Figure 4.4.

PARALLEL PROCESS
A process executed by someone (or something) else that occurs simultaneously (concurrently) with the primary process. May or may not be part of the primary process.

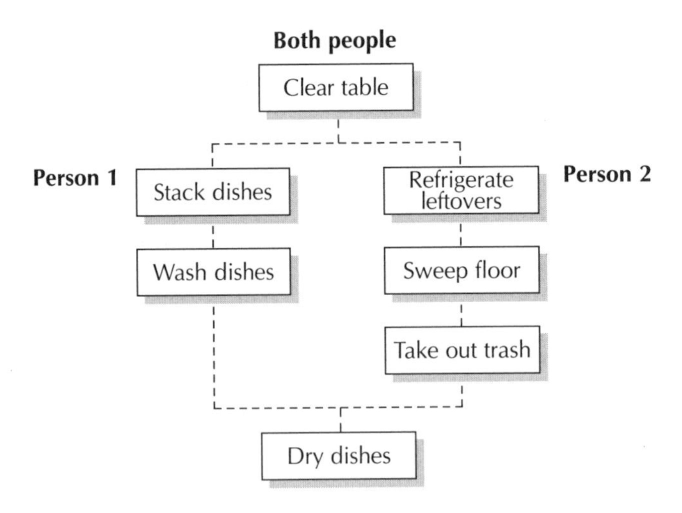

Figure 4.4. A parallel process.

CYCLE TIME
The total amount of time required to complete the process, from boundary to boundary; one measure of productivity.

Parallel processes have the advantage of reducing *cycle time*. But they normally demand more resources—either people or machines. In the previous example, an automatic dishwasher might replace one person but the tasks would need to be rearranged—since automatic dishwashers cannot clear the table or stack dishes.

We'll use rectangular, 3x5 stick-on notes to represent task rectangles. The advantage of using stick-on notes (or index cards) is that they can be easily moved around, resequenced, eliminated, reworded, or added to without redoing the whole flowchart.

Again, notice that for now we're using dotted lines to connect the boxes. Drawing the lines is one of the last steps of constructing a map, so in this book, we'll use dotted lines to show temporary, erasable pencil lines. Later, solid lines will represent the final, inked lines.

In summary

- A process consists of *steps, tasks,* or *activities* (interchangeable terms).
- Each step is depicted by a rectangle.
- Inputs (and their suppliers) are depicted by parallelograms.
- The primary process is made up of those steps that *everyone always* does.
- A parallel process is a series of steps accomplished by another, simultaneous to the primary process.
- Lines and arrows show the direction or sequence of the process.

Now, before you begin creating your own map, examine the following partially completed maps of three common processes:

☛ Setting a table

☛ Getting gas for your car

☛ Getting ready for work

You'll see these processes at various stages of completion throughout the remainder of this book. We've used these three examples throughout because they're familiar. You'll be able to direct your attention to the thought process that created the map, not the content. Answers (on page 32) to the questions give additional information not found elsewhere in the text. Therefore, we recommend that you don't skip these exercises.

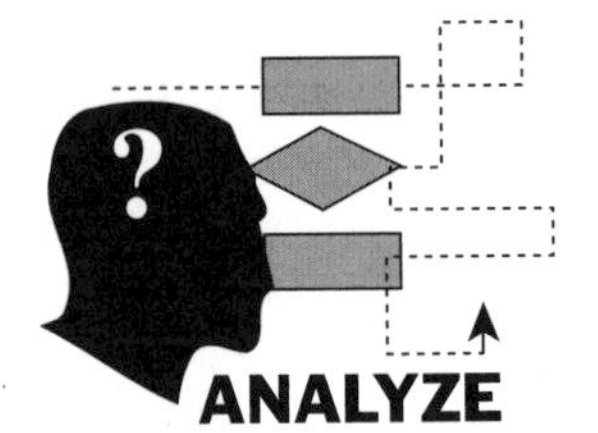

Setting a Table (primary process) 

Examine Figure 4.5 and answer the following questions.

Background: This work group listed all the things it does to set the table. One member insisted that she begins by developing a theme or color scheme. Other members thought this was fairly silly, but all it takes is one person to say he or she does something, and the item stays.

For each task the group members asked, "How many of us always do this when we set the table?" If everyone raised a hand, the item was put in one stack. If even one person said "No, I don't always do that" the item was put in a different stack.

1. How many tasks are *always* done by everyone?

....

2. What are the boundaries of the process, as shown?

....

....

3. What tasks from the right column would *you* have voted to place in the primary process?

....

....

4. Are any tasks that you do left out? Which ones?

....

....

5. What's a trivet?

....

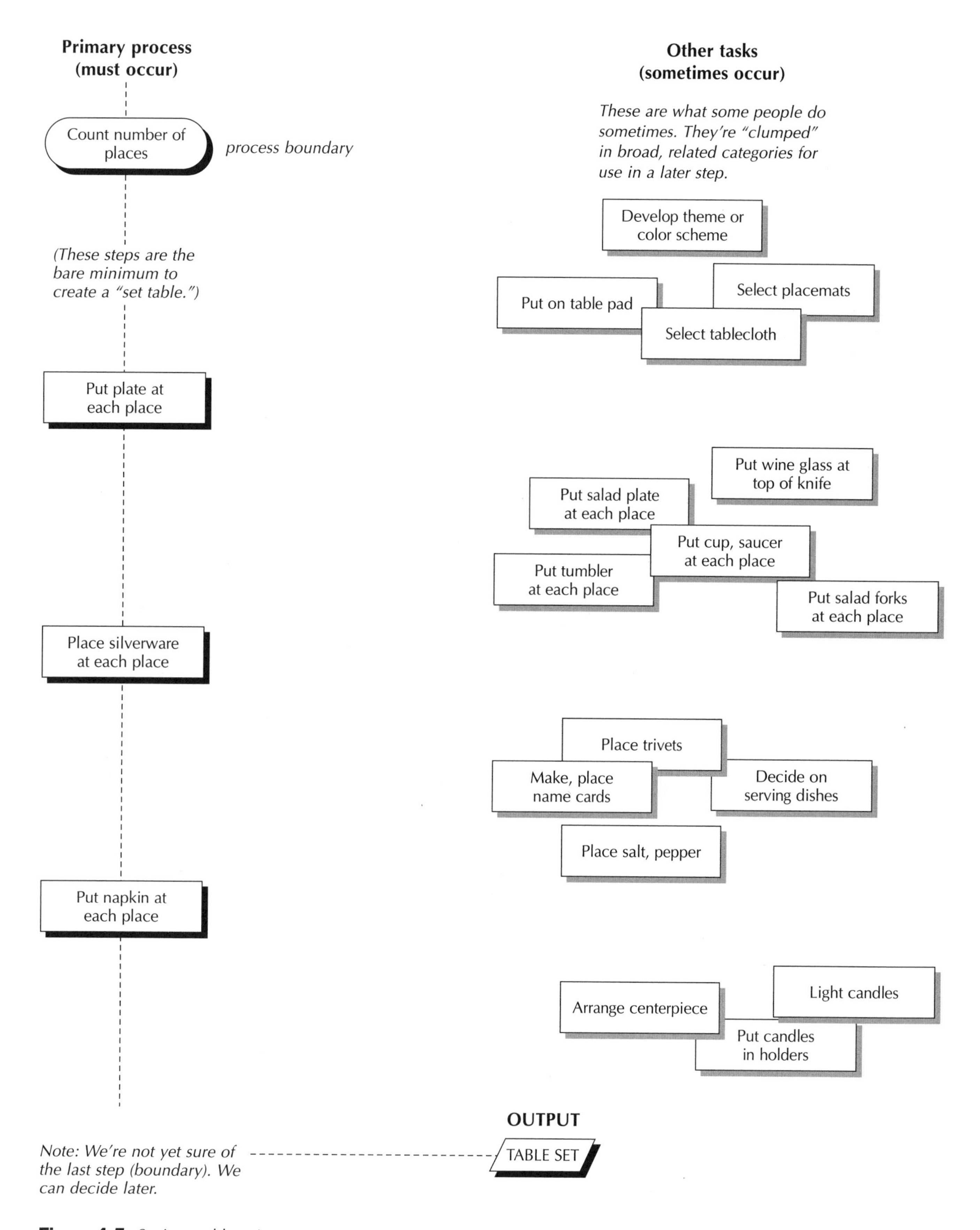

Figure 4.5. Setting a table, primary process.

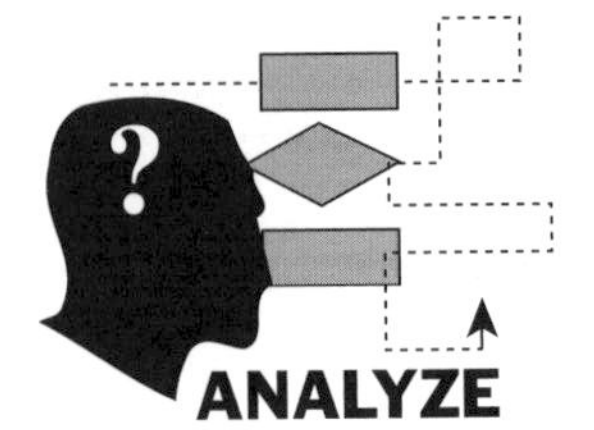

Getting Gas for Your Car (primary process) 

This process looks different. There are a lot more tasks in the primary path and fewer "sometimes occurs" tasks.

1. Where did the group members set the boundaries for this process?

....

....

2. What different boundaries might they have used (other tasks that come before or after the selected boundaries)?

....

....

3. Do you agree with the sequence of tasks? If not, how would you change it?

....

....

4. Is the section of the map beginning with the task "Remove gas cap" truly a parallel process, according to the definition? Why or why not?

....

5. Is there anything in the primary path that you never or seldom do?

....

6. Why can't we just throw out all those "other tasks" and consider the map finished?

....

....

....

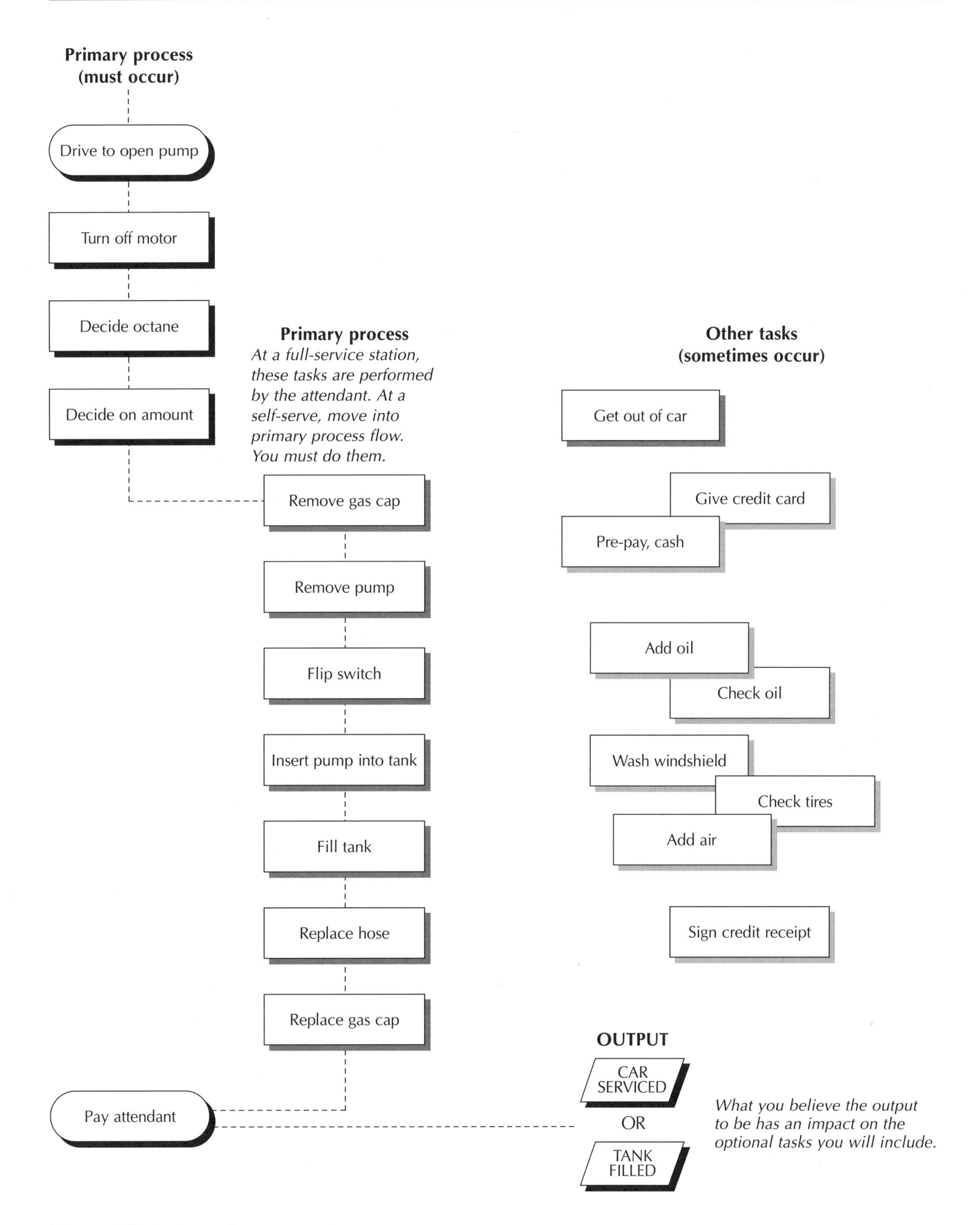

Figure 4.6. Getting gas for your car, primary process.

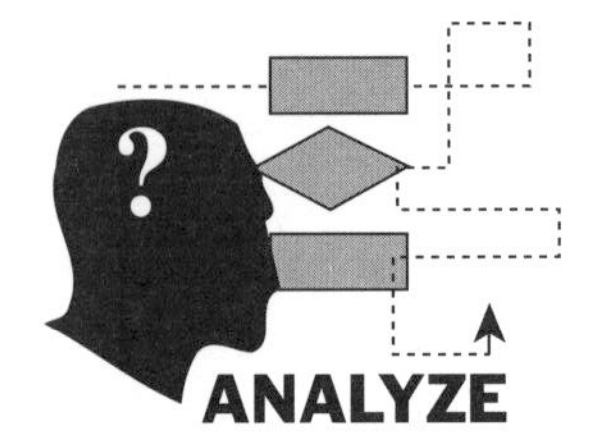

Getting Ready for Work (primary process) 

1. If someone did only the tasks in the primary path, could he or she reasonably show up for work without embarrassment?

....

....

2. Why aren't "Shower/bathe" and "Select clothes" in the primary path?

....

....

3. Circle the tasks to the right that you always do without fail.

4. Put an *"X"* through the tasks you never or seldom do.

5. Add tasks you believe were omitted or forgotten.

....

....

6. Why are "other tasks" grouped as they are?

....

....

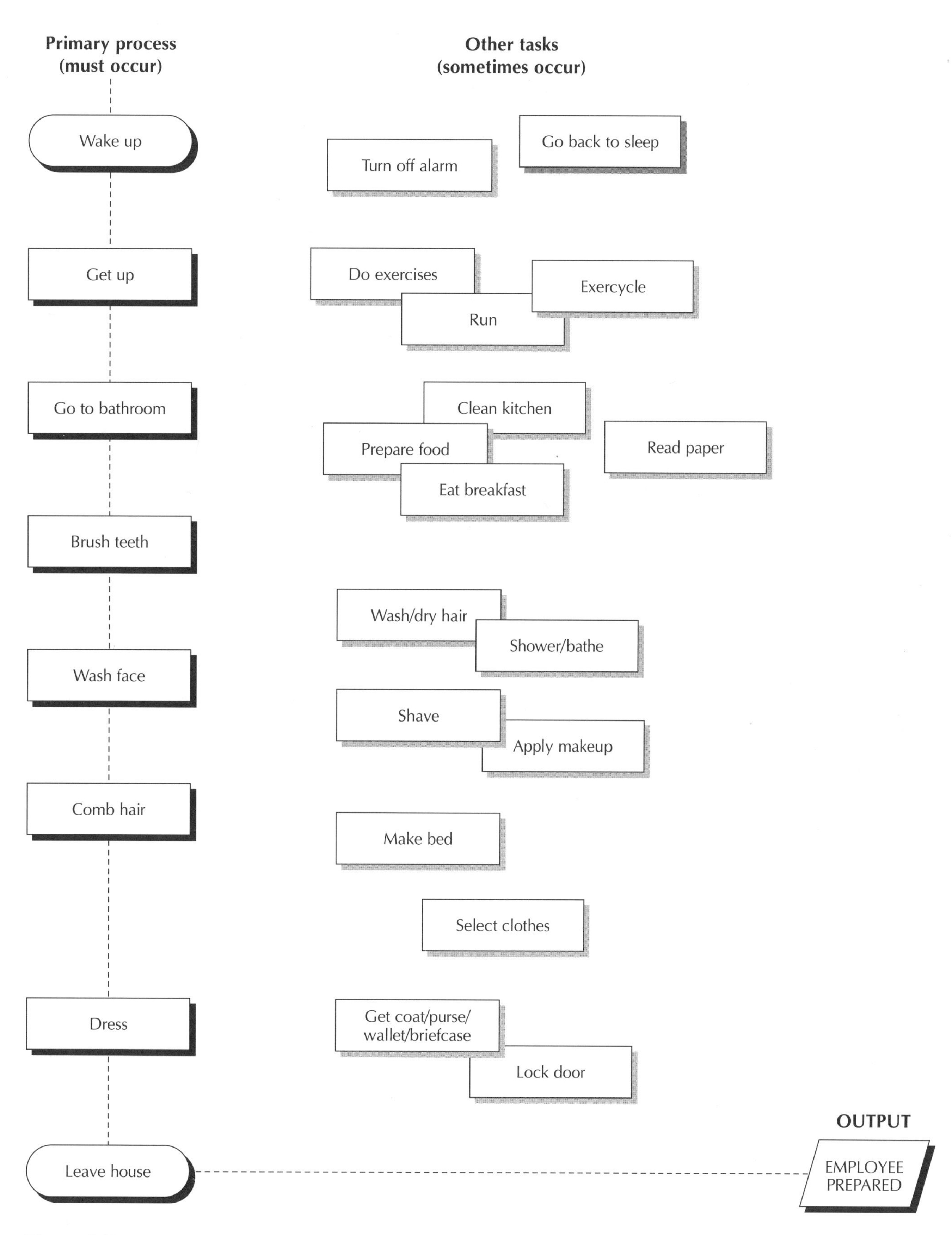

Figure 4.7. Getting ready for work, primary process.

❶ ***Brainstorm.*** On a flip chart, brainstorm a list of tasks and activities—things you do. Don't be concerned with sequence, level of detail, or accuracy at this point. Remember that brainstorming implies no judgment. You should, however, keep in mind that you are listing the "as is" rather than the "should be." Don't forget to include the boundary steps created in the previous section.

To save time, distribute rectangular, 3x5 stick-on notes and a pen to two or three people. As items are listed on the flip chart, write each task or activity on a single stick-on note.

Because you're recording what people do, each task should include a verb (such as "pay," "select," "wash") and its object (*attendant, open pump,* and so on). Thus, some typical task stick-on notes might look like those in Figure 4.8.

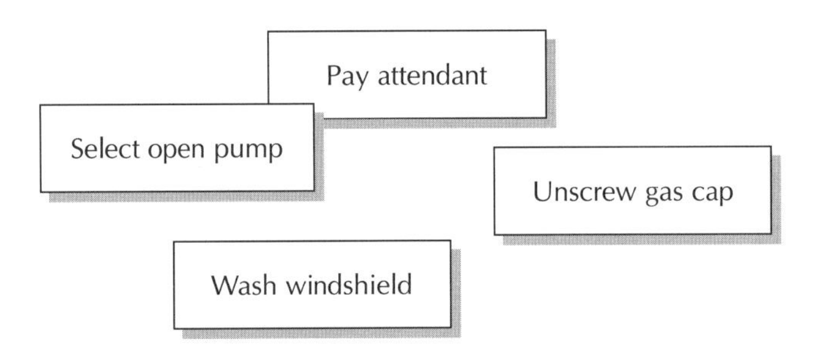

Figure 4.8. Recommended wording of steps.

Most groups will produce a list of between 30 and 60 items in 10 or 20 minutes. You'll be able to add (and subtract) tasks as you work—so don't be concerned if you don't think of everything.

❷ ***Sort.*** You'll need a large, flat surface on which to lay out the notes. A large table will work fine. All participants should position themselves to both read and move around the stick-on notes. Perform the following sorts in the order suggested.

- ☐ *Remove any tasks that have to do with "inspection," "revision," "rework," or "fix."* If the task represents an inspection that *really does occur,* set it aside. If it represents something you *ought* to do, but *don't,* discard it (you can reinstate it when you move to improvement activities).

- ☐ *Remove any tasks that may belong to another, administrative/ management process,* such as "submit travel vouchers," "attend meeting," or "prepare forecast." Most monthly, yearly, or other time-defined activities are part of some other generic process. See the list of management processes on page 9. If you can't decide/agree if the activity belongs to the process, leave it in. If it's part of another process, you'll have trouble fitting it into the flowchart later, and you can discard it then.

☐ *Examine each remaining task and place it into one of two stacks/categories:*

1. Those tasks which absolutely must occur, every time in order to produce your output (without this activity there can be no output).

2. Those tasks which occur sometimes, based on the situation or depending on personal preference (these may add value to the output, but are not absolutely essential to its creation).

The team must be unanimous to put a task in the must occur pile. The "must occur" pile represents your primary process. If you think of tasks that ought to be added to either stack, create new stick-on notes.

☐ *Discard any duplicates of tasks.*

Edit or rewrite any stick-on notes to achieve consistency of wording, either adding or subtracting clarifying words (adjectives, adverbs).

☐ *Last, if different people, departments, or functions perform some of the tasks, code the stick-on notes with a colored dot, keyed to the individual, department, or function.*

Omit this step if it doesn't apply.

❸ *Select flowchart format.*

Flowcharts may run either vertically or horizontally. Make this vertical/horizontal choice based on the amount and nature of the wall or table space available to you. You can even use a carpeted floor if you're all under 40, wearing jeans, and have no knee problems.

Tape blank flip chart pages—one after another—either vertically or horizontally as shown in Figures 4.9 and 4.10.

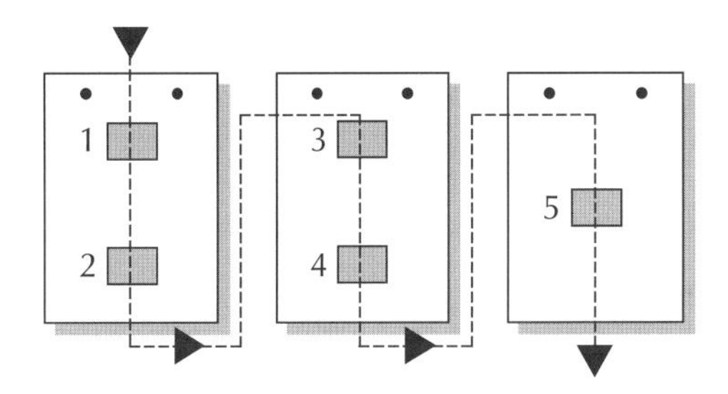

Figure 4.9. Vertical format.

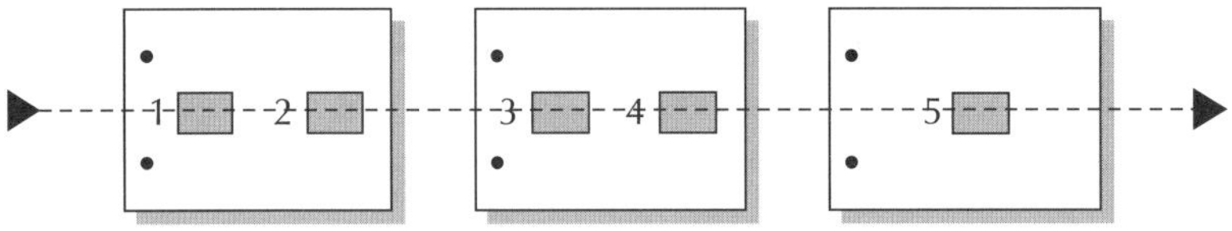

Figure 4.10. Horizontal format.

You're ready to go—unless you have a process that involves several people or offices. Macro processes tend to wander from one person, department, or function to another and back. Micro processes tend to stay put and are completed by a single person or group.

If you've selected a macro, wandering process, use the modified block format.

Modified block format

Divide each sheet into two, slightly uneven columns—the larger for your primary process, the other for all the other people, functions, or divisions as in Figure 4.11.

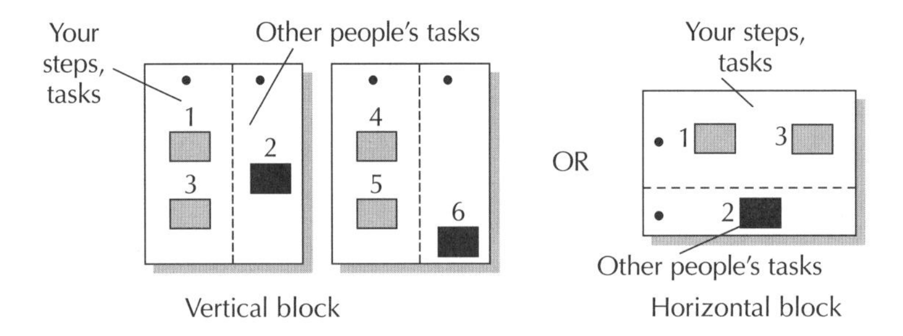

Figure 4.11. Example of modified block format.

Block diagram format

To show a macro process with several concurrent or parallel processes, you'll need pairs of flip chart pages, side by side, with one larger column for your process and a narrow column for each of the other functions/people as in Figure 4.12.

If you select the block diagram format, you will win the prize for most paper used, hands down.

❹ ***Prepare work field.***

Count the total number of task stick-on notes in your primary process. Divide by two. This is how many flip chart pages you'll need. Most groups don't believe they will use this much paper. You will.

Tape blanks firmly in place, according to your chosen plan.

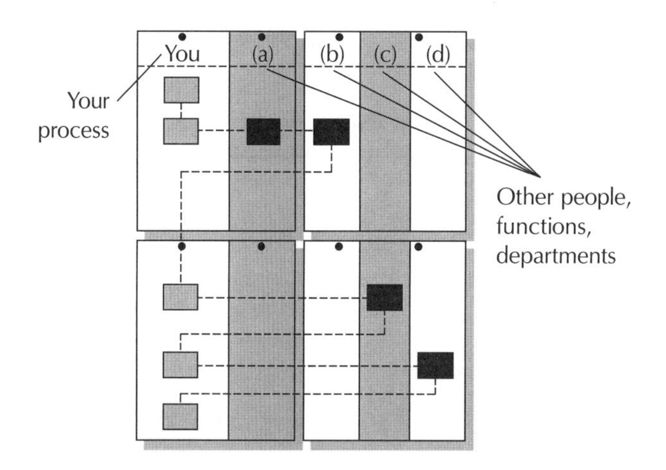

Figure 4.12. Example of block diagram format.

If you're working in a room with a large white board or chalkboard and wish to take advantage of its erasability, check first to see that the stick-on notes will stick (often they don't). You could put tape on each note to compensate. Remember too, that boards can't be rolled up and carried away easily.

❺ *Place primary process notes.*

Place the first step of the process (boundary) at the top of the first page. Place the last step at the bottom of the last page. Boundaries are typically shown as ovals. You can use another color or draw an oval with a marker to show boundaries as in Figure 4.13.

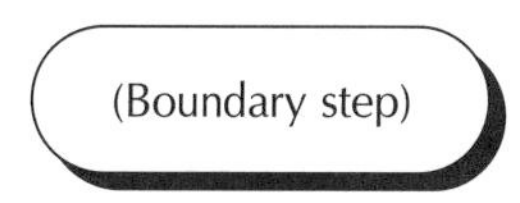

Figure 4.13. Boundary shape.

Place the remaining steps, in sequence. Place no more than two or three notes in any column, on any page. You'll be inserting lots of other things later. Do not draw any lines or arrows yet. Save the "sometimes occur" notes.

❻ *Check for reasonableness.*

Consider each step of your primary process; is it necessary to produce the output? If not, move it out of line to the stack of "sometimes occurs."

Have you forgotten any important step? Add it.

Does your work look something like the examples on the preceding pages? If not, figure out why not and make adjustments.

■

Answers to the Exercises

Setting a table (page 22)

1. *Four: count places, place plates, place silverware, and place napkins.*
2. *According to the map, the process begins by counting the places and ends with putting napkins at places.*
3. *Any task you think everybody does.*
4. *Again, your opinion. Any answer is okay.*
5. *Who cares? If you do care, it's a hotpad with little feet. The point: you have to be tolerant of other people's "crazy" ideas of what happens within a process.*

Getting gas for your car (page 24)

1. *This group chose "Drive to open pump" and "Pay attendant."*
2. *They could have started by "Driving out of driveway" or ended with "Driving out of the station." The choice of boundaries is an important one, as you map your own process. There's often disagreement among suppliers and customers as to who's supposed to do what. Setting boundaries helps clarify these misunderstandings.*
3. *Your experience may be different. Any reasonable answer is okay.*
4. *No, technically it's not a parallel process because activities are not happening simultaneously. While the attendant is pumping gas, you're probably sitting in your car doing nothing. If you got out and washed the windshield while the attendant pumped gas, you'd have a parallel process.*
5. *We can't imagine anyone who doesn't do the four steps to the left.*
6. *Because it's not very useful yet. It's the other tasks that represent the different things people do—the things you want to examine in greater detail.*

Getting ready for work (page 26)

1. *There's some room for argument here. Some might say they wouldn't be caught dead at work without having shaved, showered, or put on makeup. Others might say they could get by with just the steps in the primary process.*
2. *Because some people do these tasks at another time—such as the night before.*
3. *Any answer is okay.*
4. *Any answer is okay.*
5. *Any answer is okay.*
6. *They're grouped, more or less, in related "hunks." All the steps surrounding eating are together, and so on. This will make it somewhat easier to deal with in the next section.*

5

Map Alternative Paths

For your map to be truly useful, it must describe and allow flexibility. Not every salesperson sells exactly the same way. Managers manage differently—not necessarily better or worse, but differently. A rigid flowchart that shows a single, linear path will be discarded by those who vow not to change their successful habits for the sake of conformity alone.

ALTERNATIVE PATH
A path through a flowchart comprised of one or more optional tasks off the mandatory primary path. Preceded by a decision diamond.

The objective of this section is to build alternative paths, depending on circumstances or personal preference. Remember that you are charting your process as it is, not as it ought to be.

The symbol we will use for a decision leading to an alternative path is the diamond, depicted by a square stick-on note turned 45 degrees as in Figure 5.1.

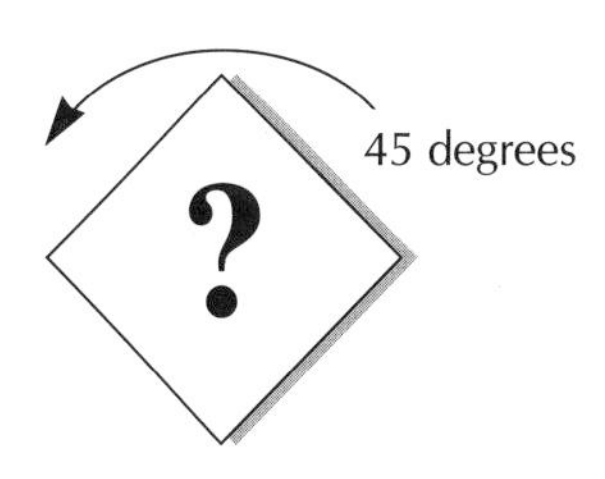

Figure 5.1. Decision diamond with question.

DECISION DIAMOND
A diamond-shaped figure that poses a question and signals either an alternative path or an inspection point.

A decision diamond always poses a question—no exception—and requires an answer. Most often, the question will lead to yes/no alternatives, as in Figure 5.2.

Figure 5.2 shows two different paths, depending on the answer to the question, "Do I need to pick up my son from baseball practice?"

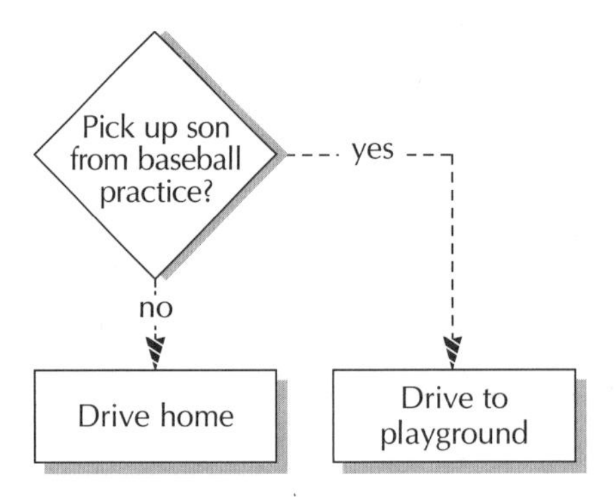

Figure 5.2. Alternative responses, paths.

Each "sometimes occurs" task demands its own decision diamond. When you go to the gas station, only sometimes do you "Get out of car." Why? What's the decision or circumstance that makes you get out of the car? Turn your answer into a question and enter it onto a decision diamond. Figure 5.3 shows the three-step thought process for creating and placing a "sometimes occurs" step with its decision diamond.

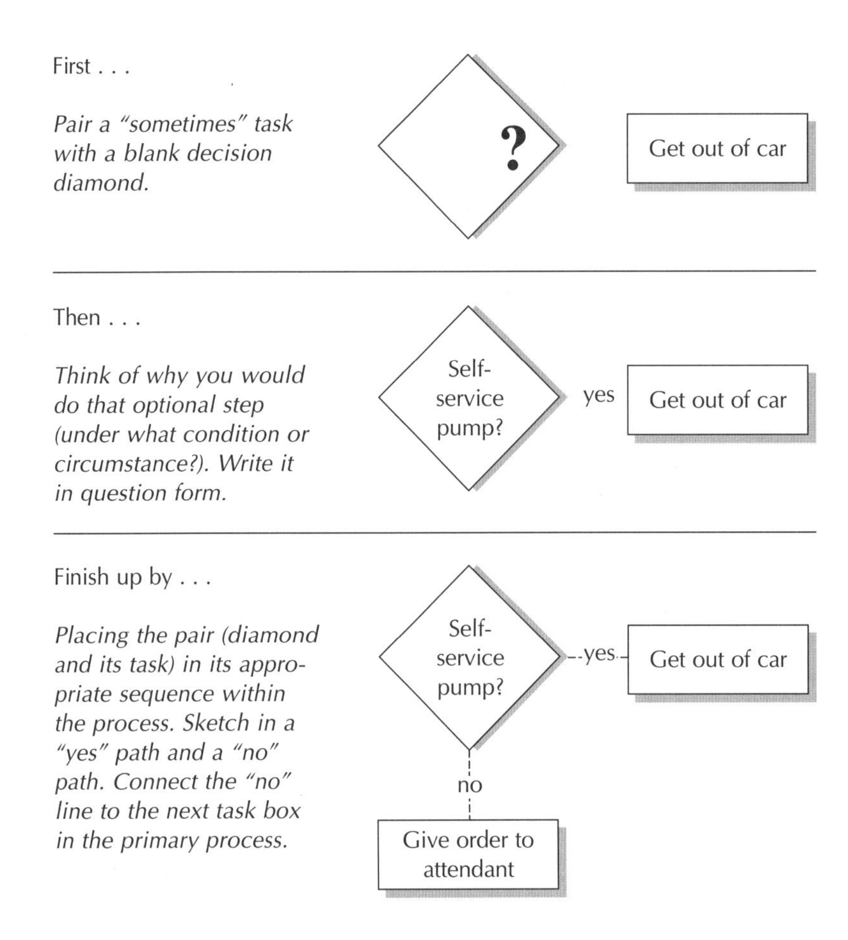

Figure 5.3. Creating a decision with yes/no alternatives.

Figure 5.4 is an example from "Getting ready for work," where the "sometimes" task is "Go back to sleep."

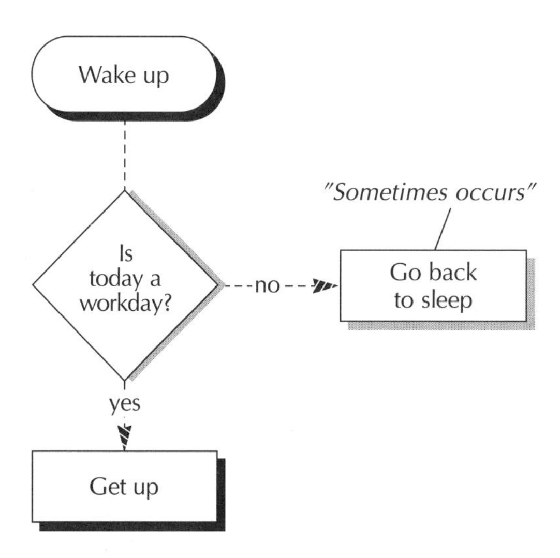

Figure 5.4. Yes/no positions reversed.

Keep the primary path running vertically, with additional loops off to either side. If you're working a horizontal orientation, keep the primary path running left to right with loops extending above and below. The "yes" and "no" paths may be reversed ("no" along primary path, "yes" off the path). It depends on how you phrase your question. Rewording the question ("Is today a holiday or weekend?") would allow you to reverse the "yes/no" directions.

One diamond, multiple paths

Decision diamonds may lead to more than two paths. Figure 5.5 shows that one of three different driving options might be chosen, depending on what after-school activity your son has.

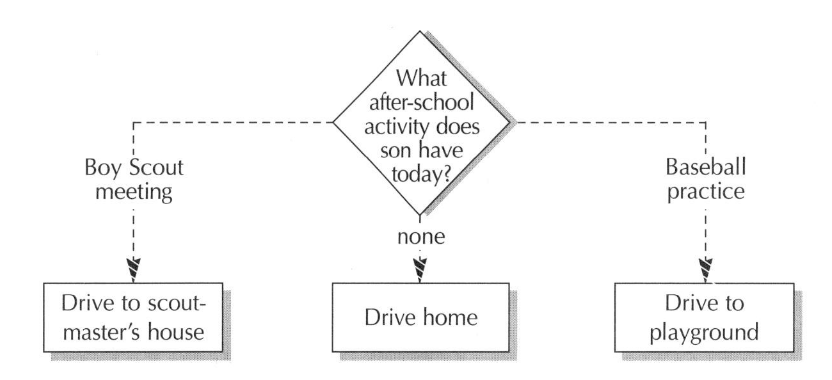

Figure 5.5. Multiple response paths.

Multiple diamonds

Sometimes there are two (or more) decisions to be made before taking an alternative path. In Figure 5.6, the "sometimes occurs" task is "eat breakfast" from the "Getting ready for work" example.

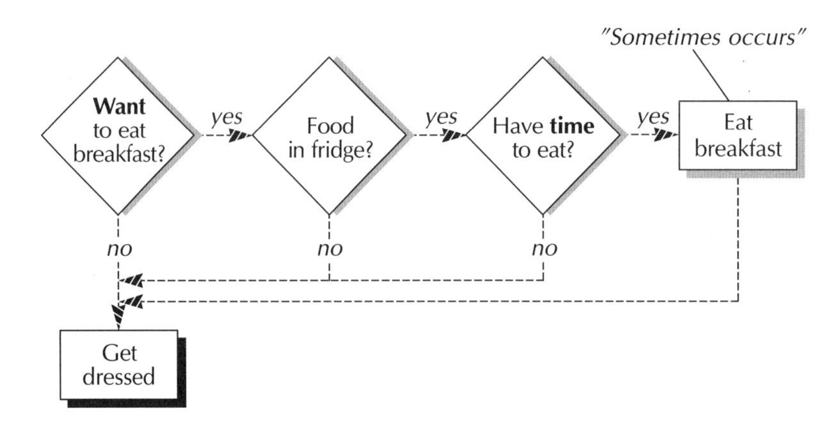

Figure 5.6. Multiple decisions.

All three conditions must be met (desire, availability of food, and enough time) before you reach the alternative step or task. A "no" response to any one of the questions eliminates the task and takes you back to the primary sequence. Theoretically anyway, you can line up any number of diamonds leading to a task off the primary path.

Specificity, objectivity

Finally, these decision questions deserve a lot of attention to how they're worded. The more specific and objective (measurable), the better. The idea is for everyone to interpret them the same way. For example, Figure 5.7 shows different ways to phrase decisions about whether to eat breakfast.

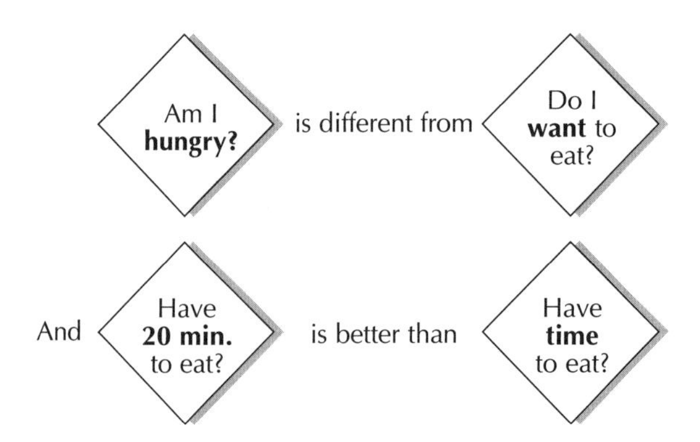

Figure 5.7. Objective versus subjective.

People will respond more consistently to the first question of each pair. In both examples, the second is open to more interpretation.

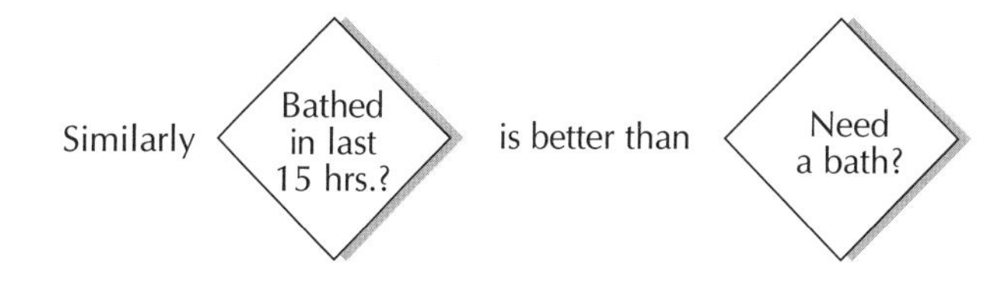

Figure 5.8. Objective versus opinion-based.

In Figure 5.8, "Need a bath?" gives people wide latitude in application. "Bathed in last 15 hours?" implies a certain standard. If people use their own judgment or act according to their subjective wishes, you could replace either with Figure 5.9.

Figure 5.9. Decision based on desire.

You'll discover, as you map your own process, that people often don't *know* why they make one decision over another. So many of the decisions in the first, "as is" version of your map may be quite subjective. Part of the later improving activities—the "should be"—will include converting some of them to a more objective, measurable state.

In summary

- Decision diamonds *must* pose a question—no exception.
- The response to a decision question may be two-state, three-state, or more. Each response creates an alternative path.
- Two or more decisions may be required to enter an alternative path.
- It's possible to respond to decision questions based on subjective judgment or an objective, measurable criterion. Generally speaking, objective is better than subjective, keeping in mind that, at this point, you're recording the "as is."
- Keep additional (alternative) tasks and loops out of the primary path; split the primary path for equal, parallel paths.
- Aim for the most flexible flowchart you have the patience to construct. This activity is the ultimate in reaching consensus. You'll make decisions later about which paths are best.

The following pages show alternative paths mapped for the three sample processes. Answer the questions, then turn to page 45 for answers.

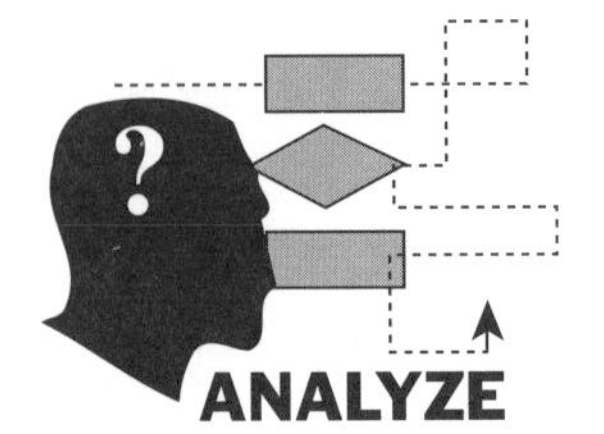

Setting a Table (alternative paths)

1. How many different situations (types of dinners) does this map account for? Name them.

   ....

   ....

2. How many (number) additional tasks are there if you're setting a table for a formal dinner?

   ....

3. According to the map, under what circumstances are placemats optional, required, or never used?

   ....

   ....

4. Theoretically, how many glasses could you put at each place?

   ....

5. The "Special occasion?" loop rejoins the primary path between which two steps?

   ....

   ....

6. If "Light candles" isn't part of the table-setting process, to what process does it belong?

   ....

7. What happened to the trivets?

   ....

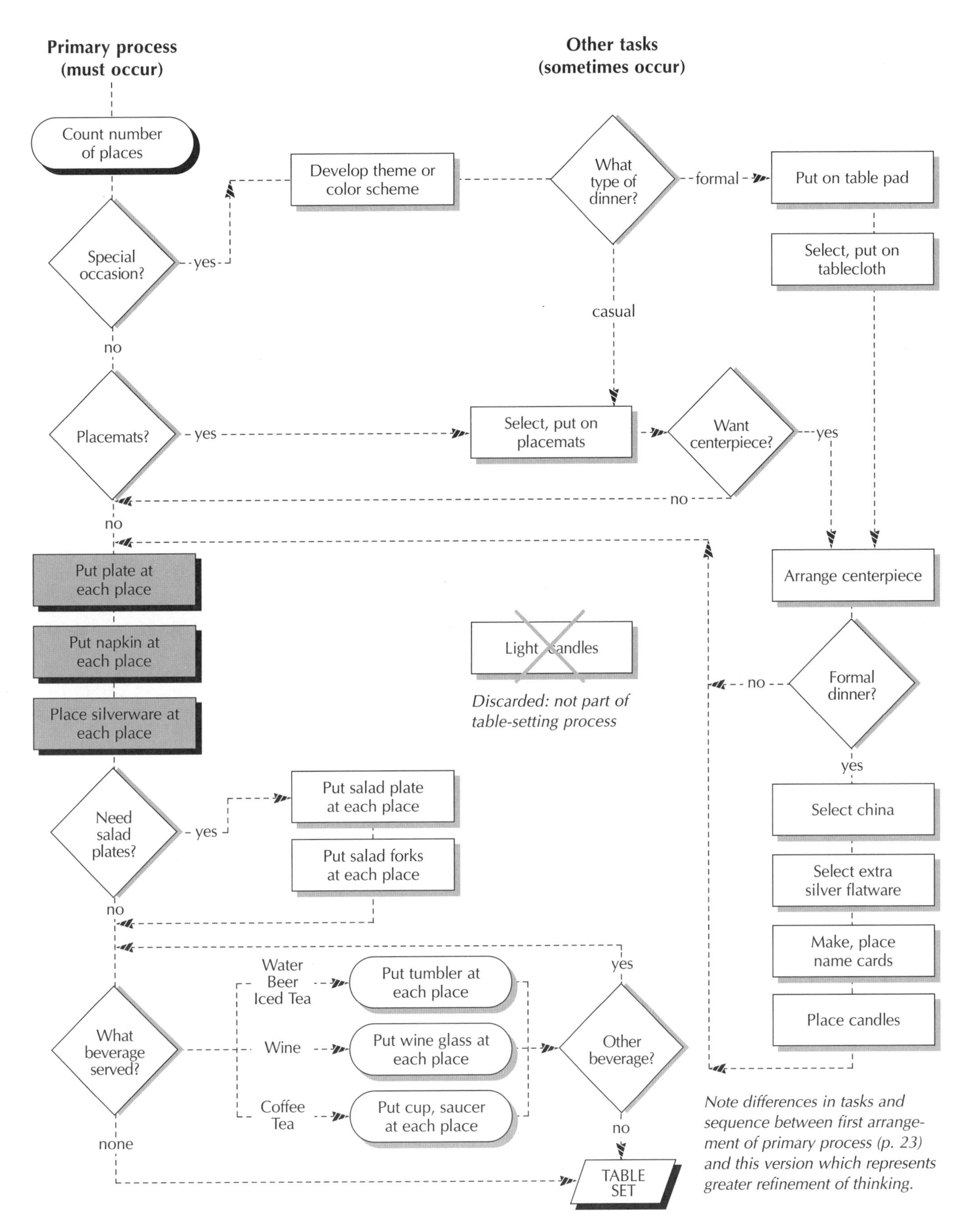

Figure 5.10. Setting a table, alternative paths.

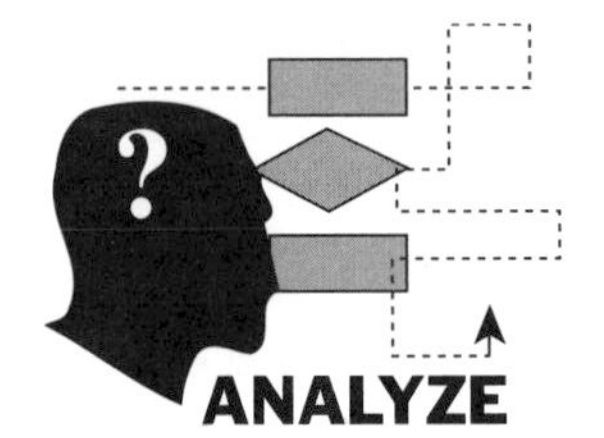

Getting Gas for Your Car (alternative paths) 

1. The boundaries of the process have changed because the group believed these were better. What was their rationale, do you think (compared with the primary process on page 25)?

....

....

2. Compared with the primary process on page 25, what other major changes to the primary process have occurred? Why do you suppose the changes were made?

....

....

3. If it were against some local law to pump your own gas, would the flowchart still be usable by that locality or would it have to be altered for people who lived there?

....

....

4. How many steps do you eliminate (for yourself) by going to a full-service pump? Does it save you any time?

....

....

5. What's the purpose of the decision diamond, "Am I in car?"

....

....

6. What are the little circles for?

....

7. Can you think of a more specific, objective way to phrase the decision question "Need oil?" Would your revised question require that the "yes" and "no" directions be switched?

....

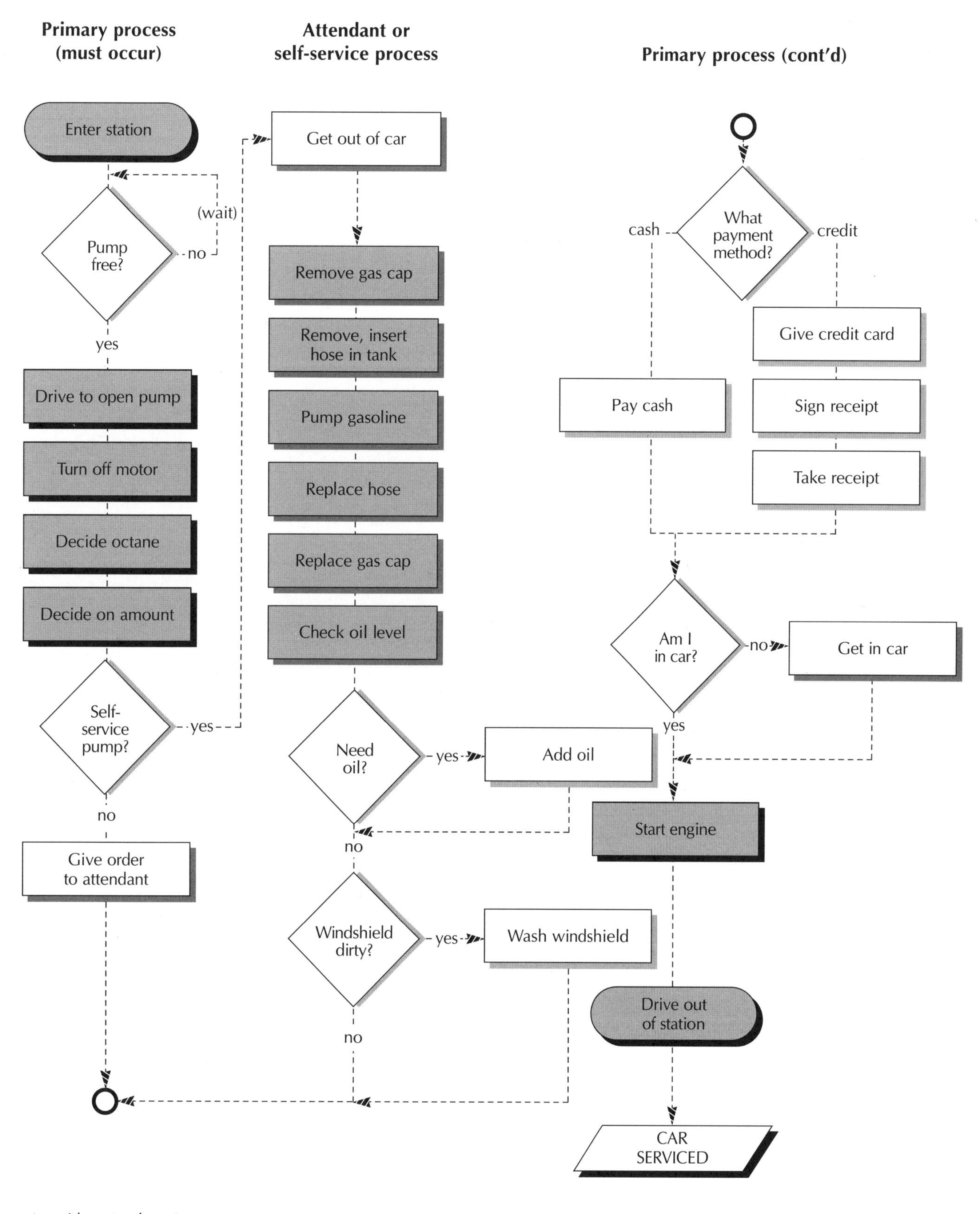

(cont'd next column)

Figure 5.11. Getting gas for your car, alternative paths.

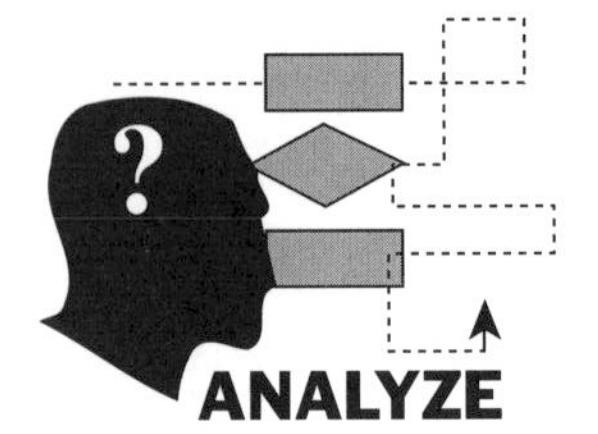

Getting Ready for Work (alternative paths)

1. What new task(s) have been added (compared to the primary process on p. 27)?

....

....

2. What two conditions must be present in order for you to exercise in the morning?

....

....

3. How many exercise alternatives are there? Could there be more?

....

....

4. According to the map, can you both shower *and* wash your face, or do you shower *or* wash your face?

....

....

5. What shaving standard is implied for men? If shaving were entirely optional, how would you reword the question in the decision diamond? Is there any standard for females and makeup?

....

....

6. How is it possible to answer "yes" to "Clothes selected?" (Under what circumstances?)

....

....

7. Suppose you had a maid or valet (or agreeable spouse). Which tasks could you give them to make a parallel process that would save you time?

....

....

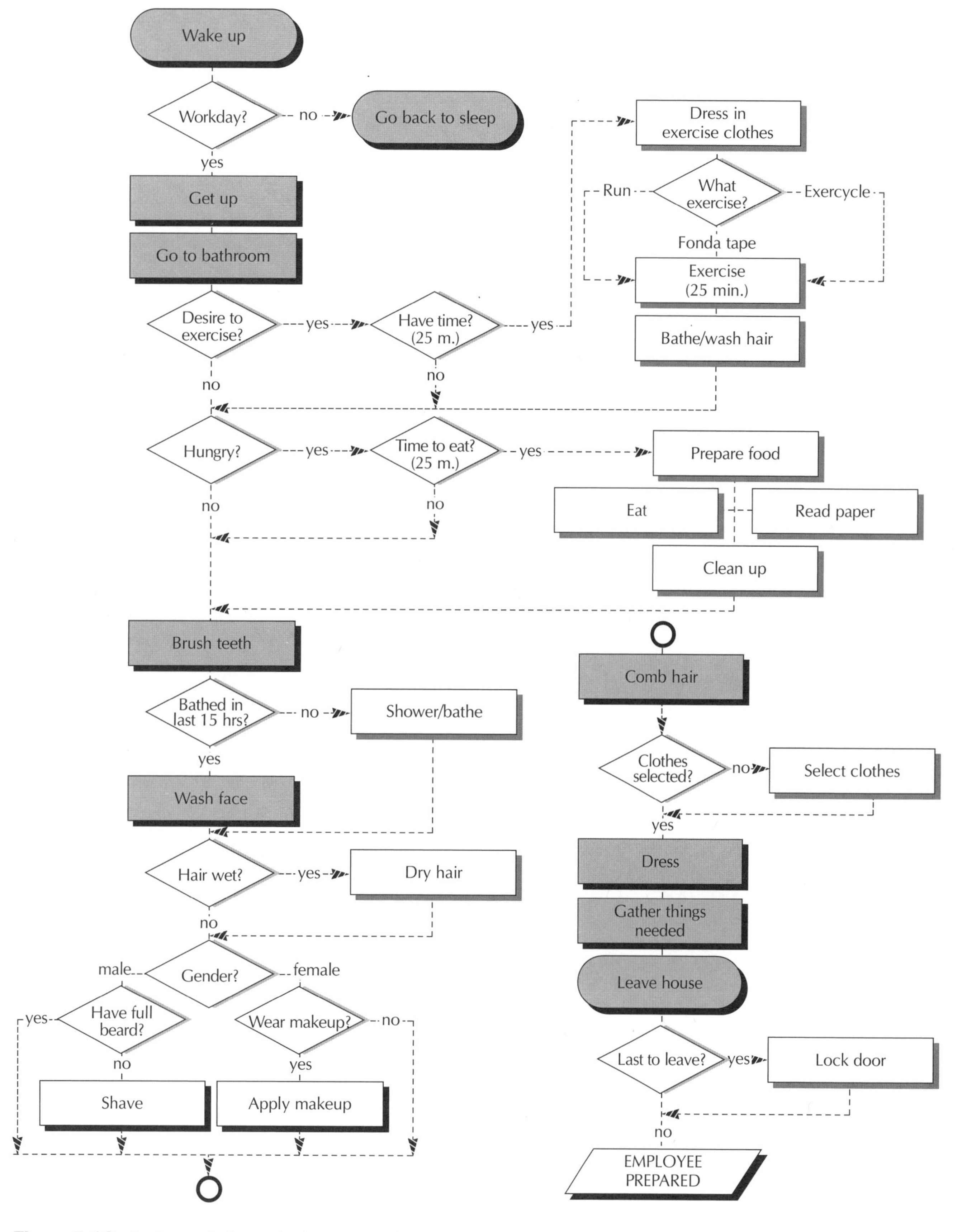

Figure 5.12. Getting ready for work, alternative paths.

❶ Make sure everyone has had time to read and understand the previous section. Discuss questions from any team member.

❷ Retrieve the stack of stick-on notes that contains tasks categorized as optional or as done only sometimes.

❸ Select one task from the stack and pair it with a square, blank stick-on note to create a decision diamond.

❹ Decide where, within the primary process, the pair (the diamond and its associated task) should be placed.

❺ Compose the question (carefully) that is to be written on the decision diamond. When you're satisfied with the wording, place the completed diamond and its associated task within the primary path in the agreed-upon sequence.

❻ Draw the answer lines (yes/no or other) lightly in pencil in the appropriate directions. You'll be moving things around, to make room for things as you go. If you use pens now, you'll have to replace flip chart sheets.

Repeat steps ❸ through ❻ for each remaining task in the stack. Add any other optional tasks or alternative paths as you think of them. This is painstaking, but it's worth it, we promise.

■

Answers to the Exercises

Setting a Table (page 38)

1. *Three. Nonspecial occasions, which is the primary process and two types of special occasions: casual and formal.*
2. *Seven. Everything from "Put on table pad" to "Place candles."*
3. *Placemats are optional for the primary process, required for special occasion casual, and never used for a formal dinner.*
4. *One for every beverage you intend to serve; theoretically, every glass and cup you own!*
5. *Between the first boundary, "Count number of places" and the second step, "Put plate at each place.*
6. *Probably a step in the "Serving dinner" process.*
7. *The person who suggested the step in the first place admitted she was just showing off and agreed to eliminate it from the process.*

Getting Gas for Your Car (page 40)

1. *Someone in the group pointed out that cycle time could be affected by waiting for an open pump. Therefore the first boundary should be "Enter Station." The process actually ends when you drive away—not when you pay the attendant.*
2. *To keep the level of detail about the same throughout the entire process, some steps were eliminated in the "Pump gas" sequence.*
3. *The flowchart would still work fine. People who live in such localities would always answer "no" to the question, "Self-service pump?" and would always answer "yes" to "Am I in car?"*
4. *You save yourself seven steps for sure and possibly two optional steps ("Add oil" and "Wash windows"). You probably save no time because you're sitting in the car waiting for someone else to complete the steps. (This is why you can't call those steps a parallel process.)*
5. *It accounts for the person who has gotten out of the car at a self-service pump.*
6. *The circles are a handy little convention to show the continuation of the process. Use them at the bottom and top of pages with the page number inside.*
7. *"Dipstick reads at least 'one quart low'?" is more specific than "Need oil?" Yes/no paths would stay the same. But if you phrased the question "Dipstick reads 'full'?" you would have to switch the yes/no paths. Some people are sticklers about keeping the yes/no paths consistent. We don't think it makes a bit of difference.*

Getting Ready for Work (page 42)

1. *"Gather things needed"(coat, briefcase, purse, lunch, and so on) has been added. The point: as you see things to change, do so. Your first efforts often overlook important things.*
2. *You must want to exercise, and you must have 25 minutes to spend.*
3. *This map shows three alternatives. The possibilities are almost limitless. The difficulty comes in finding room for them on the map.*
4. *If you shower, you reenter the primary process after the "Wash face" step. Thus you either shower or wash your face, not both.*
5. *Men without full beards must shave. To make shaving optional, the decision diamond could read "Feel like shaving?" The implication for females is that they must be consistent. Either they always wear makeup or never wear it. As you develop your own paths, be sure to say what you mean. Small differences in wording make big differences in meaning.*
6. *If you did it the night before—not part of this process.*
7. *"Prepare food," "Clean up," "Select clothes," and "Gather things needed" are the only possibilities. While the maid or valet could help you brush your teeth or dress, you still must be involved. So no time is saved if you're involved too.*

6

Map Inspection Points

Inspection points serve to *find* errors before they reach the customer. A later step in improvement attempts to *prevent* errors. However, as part of mapping the "as is," you'll need to understand how errors occur in your existing process. Figure 6.1 shows errors detected by customers after delivery of the service or product, a costly proposition.

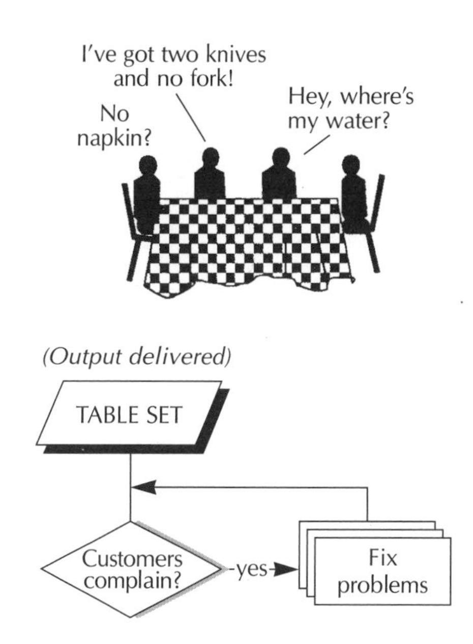

Figure 6.1. Error detection after delivery.

INSPECTION POINT
A pass/fail decision, based on objective standards, to test an output in process. Signaled by a decision diamond with two or more paths leading from it. May lead to a rework loop (step) or to a do-over loop.

An inspection point is a special category of decision diamond that typically leads to a "pass/fail" answer. Work that fails inspection causes the process direction to reverse itself. The flow progresses from top to bottom, but a "fail" forces the direction back up the process and adds a "rework" task as shown in Figure 6.2.

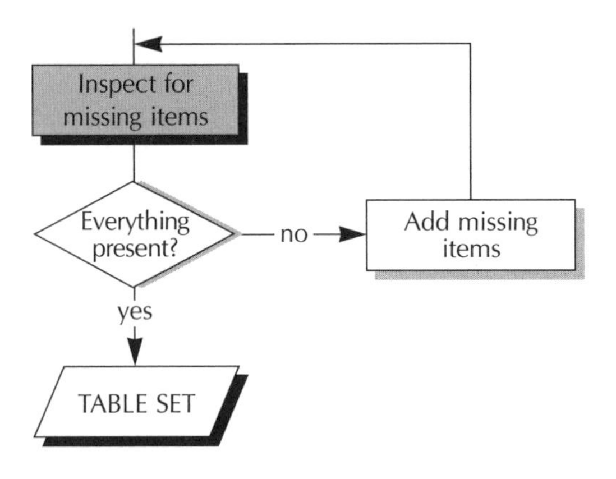

Figure 6.2. End-of-process error detection (before delivery).

REWORK LOOP
The result of a failed inspection point. A rework loop adds steps to the process and generally leads back to the inspection diamond.

The "rework" loop in Figure 6.2 may be circled several times—until the table passes inspection. The good news is that the inspection catches errors before customers do. The bad news is that rework loops add to cycle time and, ultimately, to cost.

When the error is bad enough, rework isn't sufficient; you must scrap the output and start over as shown in Figure 6.3.

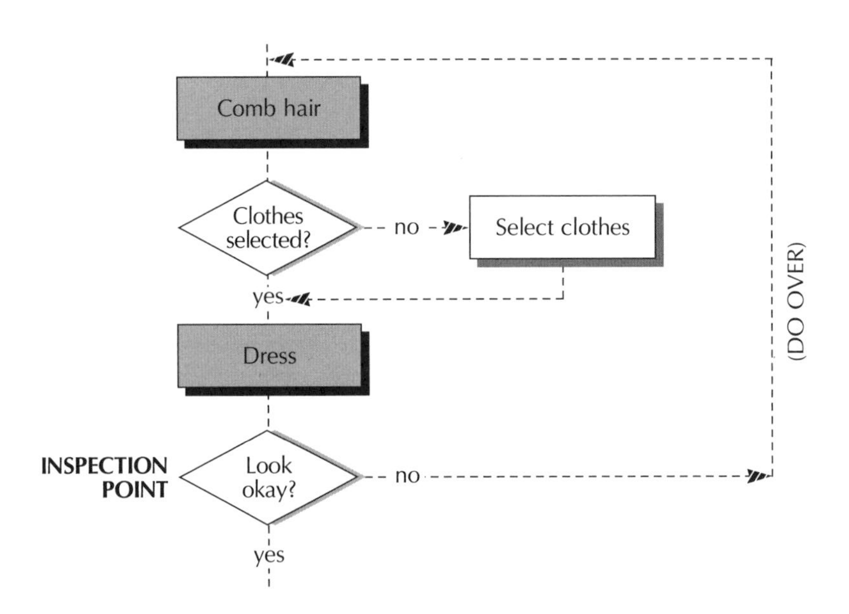

Figure 6.3. Do-over loop.

DO-OVER LOOP
Another result of a failed inspection point, a do-over loop leads to an earlier step in the process. Steps must be repeated. Associated with scrap.

Do-over loops deliver the same good news/bad news dilemma as rework loops—it's better to find them before the customer does, but they add time and cost.

Inspection points may be formalized—staffed with inspectors and checklists, or they may be rather laissez-faire arrangements where someone happens to notice a problem and sends it back. Both types may be mapped the same way.

Specificity, objectivity

If we're inspecting gizmos in a manufacturing process, it's clear that an inspection point that says "Look okay?" will yield sloppy results. Giving the exact dimensions clarifies what you mean by "okay" (see Figure 6.4). Everyone will interpret specific inspection questions the same way.

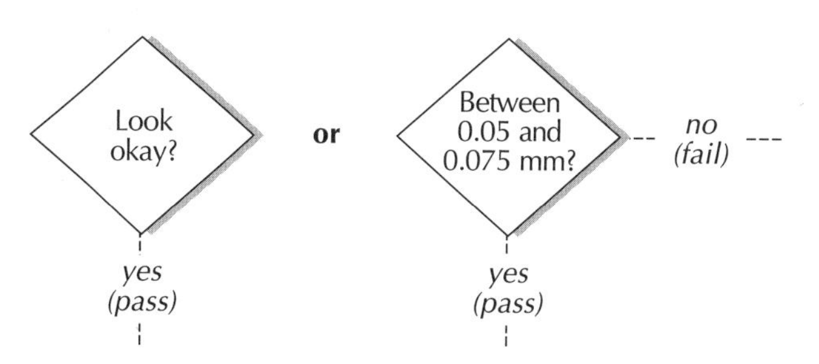

Figure 6.4. Subjective versus objective inspection criteria.

And yet, when we're dealing with processes that produce less tangible products, we're tempted to accept these broad, subjective "look okay?" kind of standards.

Suppose, as in Figure 6.5, we use a vague "look okay?" criterion to determine whether we need to do-over or rework our appearance.

Ask everyone in the group how they decide if they "look okay" and you'll very likely get dozens of different ideas—or worse—no ideas ("I just know, that's all").

At this point in your mapping activity, you can decide to accept a subjective inspection point if you're convinced it represents the "as is." Mark the inspection step card or stick-on note with a big star or border it in red. This means "come back and fix this." For instructions on how to fix it, read the section "Develop and Apply Standards" on page 63.

Otherwise, spend some time trying to uncover the existing objective (measurable, observable) criterion or standard.

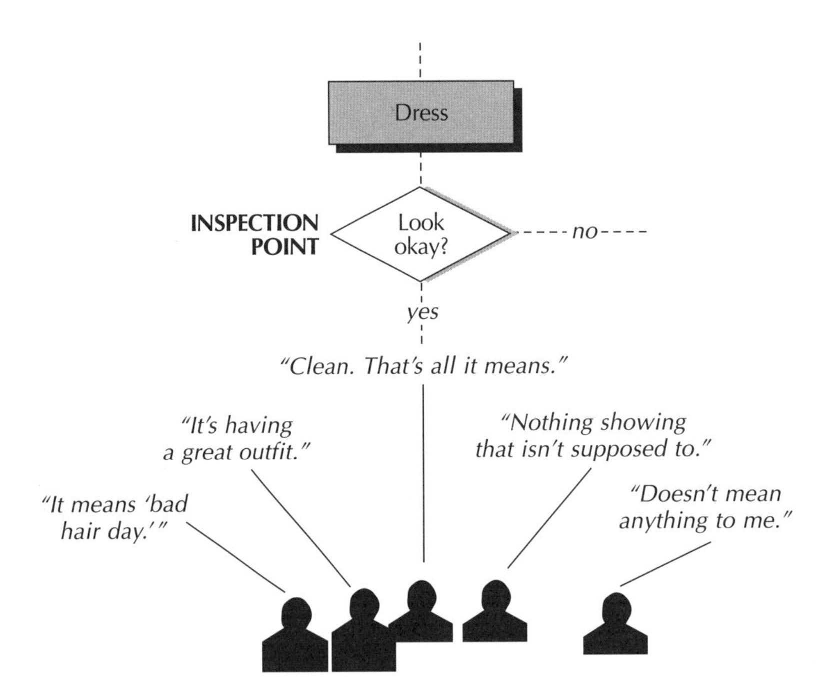

Figure 6.5. Interpreting subjective criteria.

In summary

- Inspection points uncover errors and flaws—either formally or informally.
- Inspection decision points pose a pass/fail question. A fail reverses the flow (back toward the beginning of the process).
- The reversed-direction loop creates either a rework step or a do-over loop.
- Rework loops require one or more additional steps and rejoin the primary process above the inspection step/diamond.
- Do-over loops (scrap) lead back to an earlier step in the primary process which will be repeated.
- Inspection points represent standards. As such, they should be specific, objective, and measurable. If they are not, this is an area to improve.

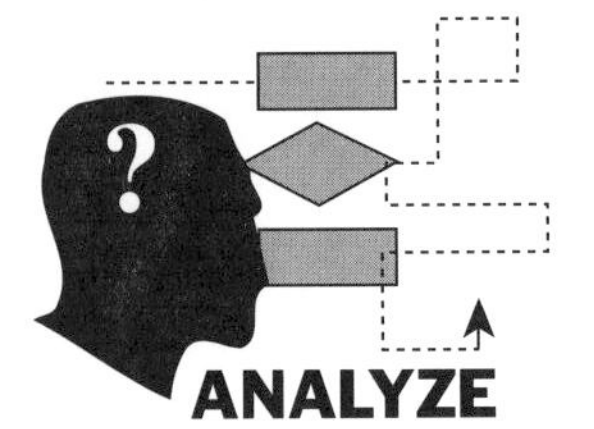

Setting a Table (inspection points) 

Our table-setting team members began by asking themselves, "What errors are common in the table-setting process?" Their short list: 1) Too many or too few settings, 2) Too formal or too casual a setting, and 3) Incomplete setting(s).

1. The first error, Too many/few settings, occurs early in the process—between "Count number of places" and "Put plate at each place." Why did the team choose to change the wording of the boundary task from "Count number of places" to "Count number of guests?" What difference does it make?

....

....

2. The first inspection point (black diamond) says that, after the plates are placed, they are counted. If that number doesn't match the number of guests, plates should be added or subtracted. Can you think of a more efficient way of inspecting for (or avoiding) this error?

....

....

3. Does inspection at point 1 occur 100 percent of the time, or on a random basis?

....

4. No inspection point was mapped for error 2. Why?

....

....

5. If you were in charge of "Table inspection for complete settings" (point 3), how would you (do you) do it? What's the mental process?

....

....

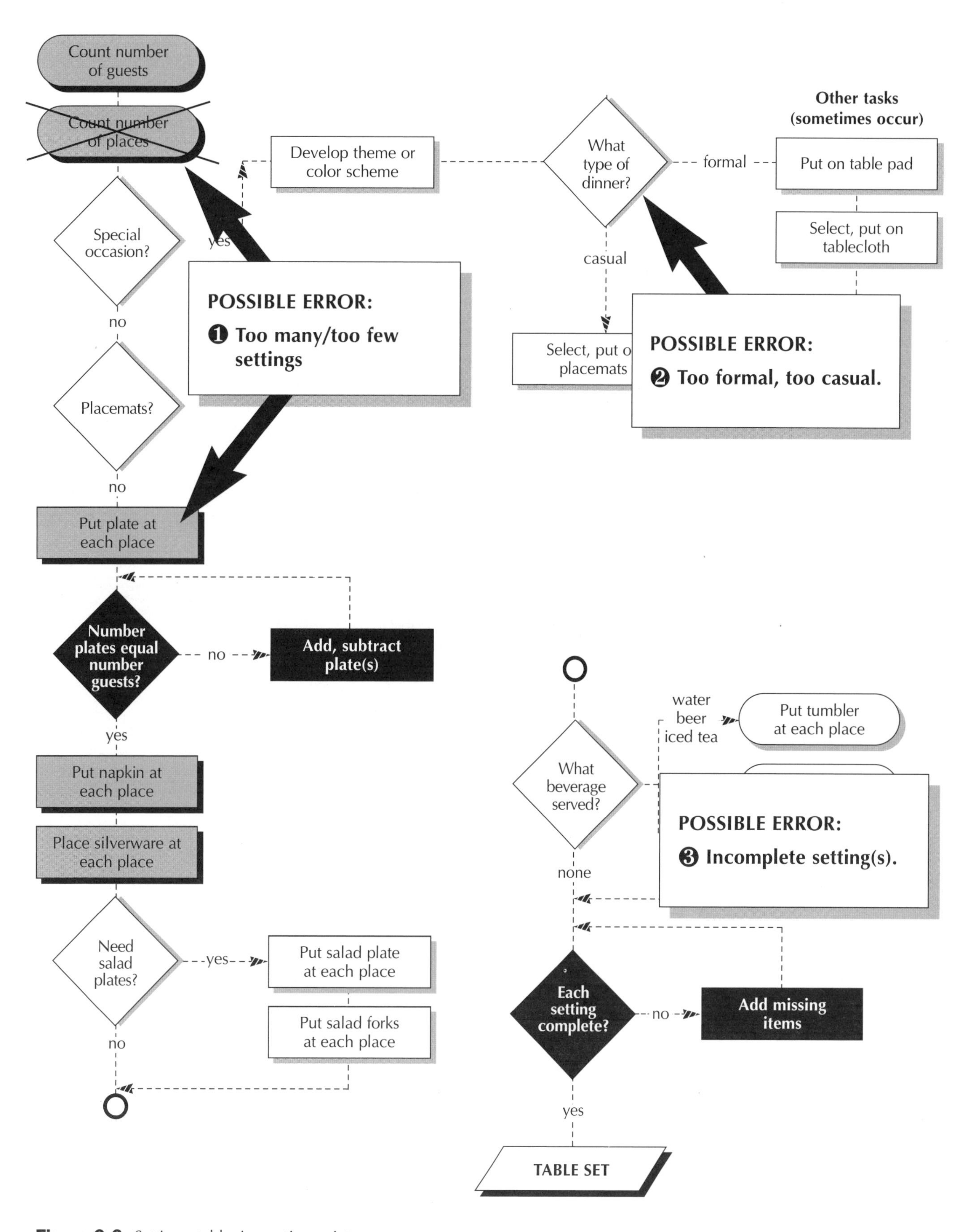

Figure 6.6. Setting a table, inspection points.

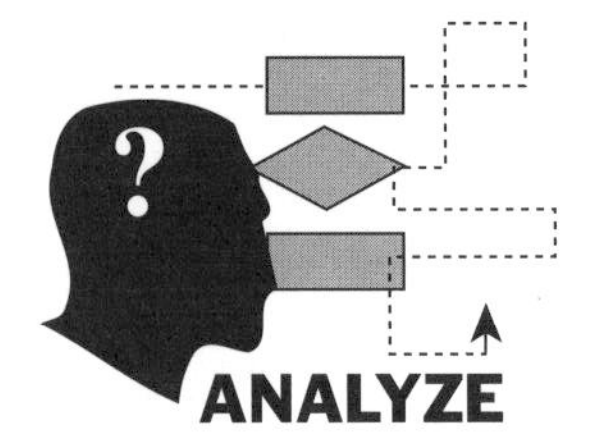

Getting Gas for Your Car (inspection points)

1. What two errors resulted in the choice of inspection points?

....

....

2. If you drive to a pump and find that the hose won't reach, what steps must you take to correct the situation, according to the map?

....

....

3. Leaving the inspection diamond where it is, can you think of another action to take (other than "Select different pump") that would shorten the do-over loop?

....

....

4. How effective do you think the inspection point "Gas cap in place?" will be in catching the gas cap error before driving away from the station? Why?

....

....

5. It's very hard to avoid thinking about ways to improve a process when you see inspection points like the two in this example. Why should you document the "as is" rather than the "what should be?"

....

....

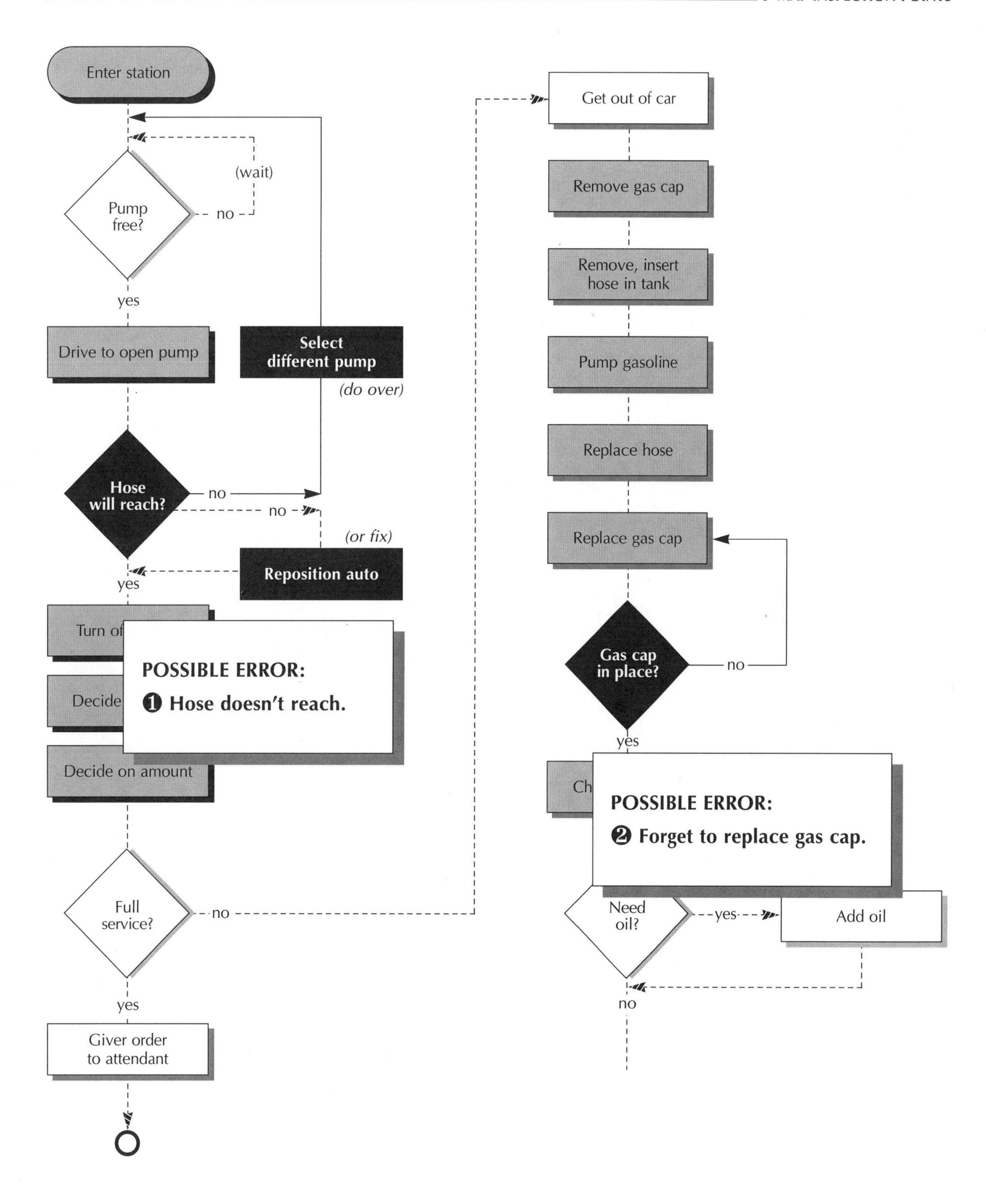

Figure 6.7. Getting gas for your car, inspection points.

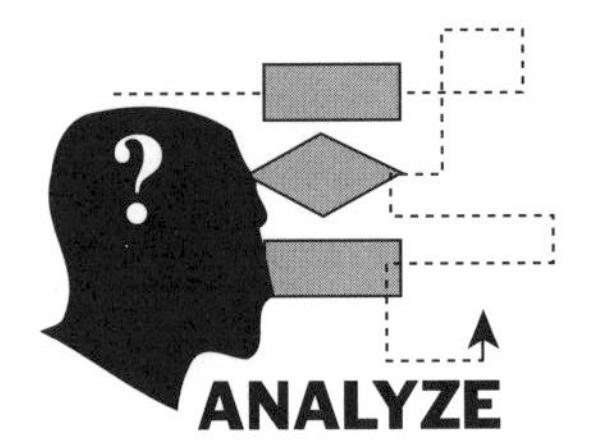

Getting Ready for Work (inspection points)

1. Suppose you determined that you did not "look okay" because you had checked "no" for "Hair neat?" What steps does the map direct you to take?

....

....

2. What steps does the map direct you to take if you checked "no" for "Clothes clean?"

....

....

3. The "Have everything?" inspection point is an aggregate of how many different, separate inspection diamonds?

....

....

4. Are any of the items on the "Look okay?" checklist too vague to be applied the same way by everyone? Which ones?

....

....

5. How important would it be for everyone to agree on the checklist for "Look okay?" How do you decide?

....

....

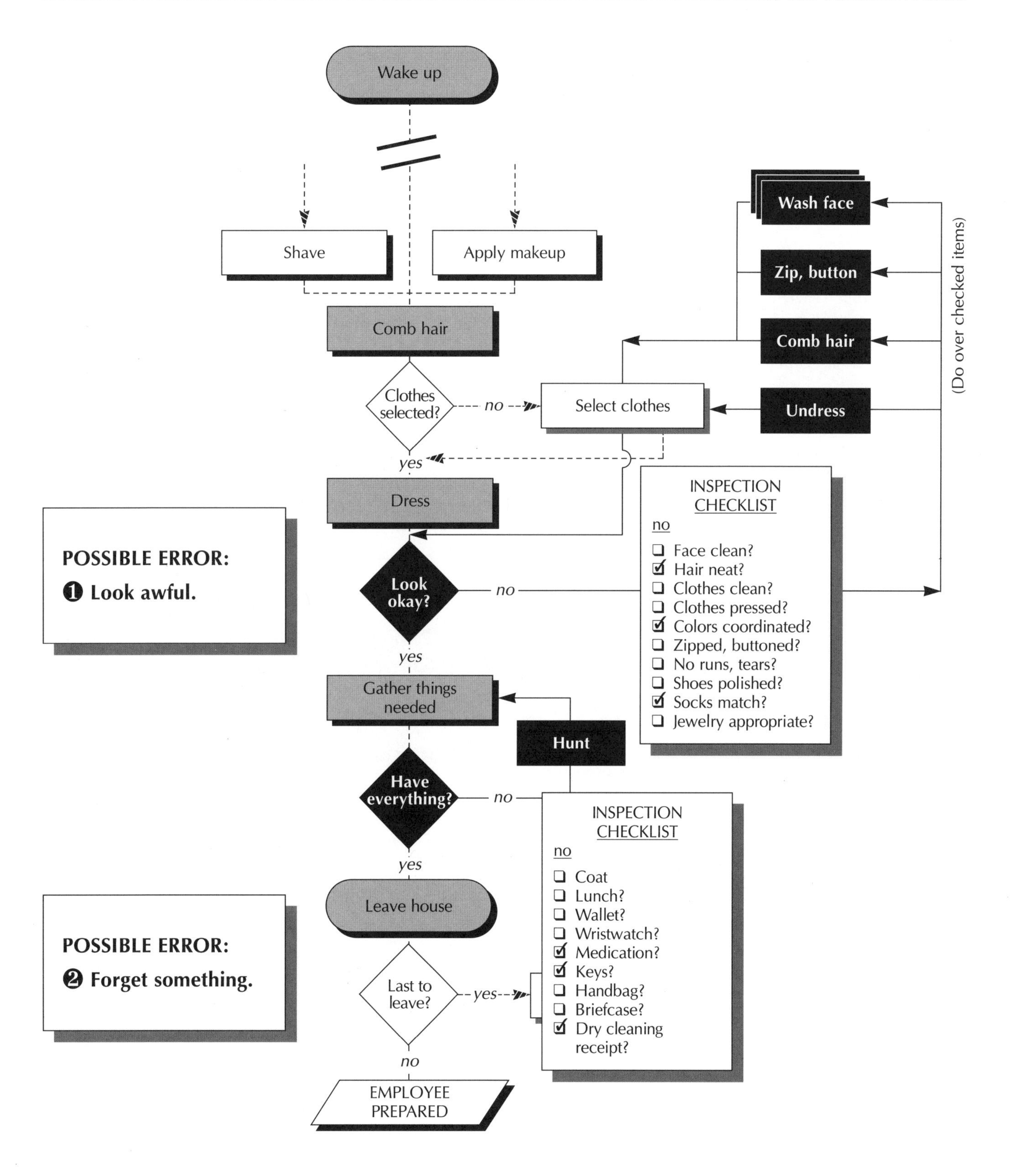

Figure 6.8. Getting ready for work, inspection points.

❶ *Make sure everyone has had time to read and understand the previous section.* Discuss questions from any team member.

❷ *Make a list of likely errors. Start with any discarded stick-on notes that represent "check," "approve," or "inspect for errors." Reduce the list to those that you actually inspect now. Save the list for use during improvement activities (the "should be").*

❸ *Locate the first inspection point on the flowchart. Identify the person who performs the inspection*—the process participant, the manager, or one of the other functions/columns in the block diagram.

❹ *Create (or find) the stick-on note that describes the inspection task* (such as review draft, inspect subassembly, or review sales plan).

❺ *Create a decision diamond that represents the pass/fail decision. Place the pair of stick-on notes in the appropriate sequence of the process.*

❻ *Chart the revision loop, adding necessary rework steps. Or, chart the do-over loop.*

Repeat steps ❸ *through* ❻ *for remaining inspection points.* If you inspect the final product before it is delivered to the customer, show that at the bottom of the process. This may serve to change your process boundary.

Finish the flowchart.

❼ *Talk through every path, making certain the elements are in the correct place and sequence.* Make any changes necessary.

Keep the spacing loose—much looser than we show in the examples.

❽ *Draw the connecting lines with a pen, finally.* Depending on the mess you've made, you may have to replace flip chart pages.

❾ *Make a record of your map.* Use a Polaroid camera for a quick and easy way of making a copy. Or, have someone redraw the map on a piece of paper or with a computer drawing program. You'll need to know where you started as you begin to make changes and improve the process.

■

Answers to the Exercises

Setting a Table (page 52)

1. *It's the number of guests that determines the number of places to set. With that small change, the inspection point is easier to express. The point: you'll undoubtedly find a number of changes to make as you proceed. You should feel absolutely free to change previous work—don't consider your map as carved in stone.*
2. *Don't wait until the plates are on the table to count. Count them as you take them off the shelf in the kitchen.*
3. *This one probably requires a 100-percent check, unless you can think of a way to avoid the error altogether (in the subsequent improvement stage).*
4. *The error at this point probably occurs because of a misunderstanding of the terms casual and formal. Your definition of formal and mine may be considerably different. To avoid the error, you don't need an inspection as much as you need commonly understood definitions. Always ask, "What do you mean by . . . ?"*
5. *Most people (with experience) will have a mental image of what a complete setting looks like. To help others with less experience, it's useful to have the experienced person devise a written checklist that others can follow.*

Note: Can you see now that we're unable to fit the entire process on a single page? It just keeps getting bigger and bigger. And we've got farther to go.

Getting Gas for Your Car (page 54)

1. *Finding that the hose doesn't reach the tank and forgetting to replace the gas cap.*
2. *As the map reads, you must find another pump (presumably pumping from the other side), wait for it to free up, then drive to it.*
3. *You could reposition the car closer to the pump and stretch the hose.*
4. *If you're forgetting the step "Replace gas cap," you're just as likely to forget to inspect. This kind of inspection point is wishful thinking unless there is some means to remind you to do it.*
5. *Both inspection points in this process are fairly silly. If the map really reflected the "as is" process, we'd have to admit that we didn't actually do any inspection. Maybe we should, maybe not. Check yourselves often about whether you're mapping what you think ought to be rather than what is.*

Getting Ready for Work (page 56)

1. *Comb hair, then go through the inspection diamond again. By now you'll have a cobweb of lines that begin to cross. Note the half circle that shows the solid line jumping over the dotted line beneath it.*
2. *This is a particularly time-consuming error because the step "Undress" has been added. Then you must select a new set of clothes, dress, and submit yourself to another inspection.*
3. *Nine. The checklist takes the place of nine separate decision diamonds. Checklists tend to keep things simpler, graphically.*
4. *Yes. "Hair neat?" "Colors coordinated?" and "Jewelry appropriate?" are all open to interpretation and different values. They're worth discussing, but don't expect quick agreement.*
5. *This is the very heart of the standards issue. While it's unlikely any rational person would insist on dress standards (we went through that in the 1960s and 1970s), there are other processes where standards are both appropriate and necessary.*

7

Use the Map to Improve the Process

So you have a fairly good picture of how you do your job now. But what if you could give your customers an error-free output every time? What if they could have that output without waiting? What if you could offer this instant, perfect product or service at a lower cost? You'd have a thoroughly delighted customer and a quality product or service. It's an achievable, though not instant, goal.

Broadly speaking, the 13 improvement techniques in this chapter and the next affect either error rates or cycle time; some of them operate on both, simultaneously. By improving error rates and cycle times, you directly or indirectly affect both cost *and* customer satisfaction.

Read through all the techniques before you begin improvement work. The five techniques in this chapter should always be applied to a process. The remaining eight are optional but recommended. And remember that process improvement is a continuous activity. You can't do everything at once. It's better to work at a reasonable pace over a longer period than to try to improve everything all at once. Thoughtful improvements take time for ideas to develop. Don't rush or set unreasonable deadlines.

Techniques in this chapter

1. Eliminate or minimize nonvalue-added steps.
2. Develop and apply standards.
3. Move inspection points forward.
4. Eliminate the need for inspection points altogether.
5. Chart and evaluate inputs and suppliers.

1

Eliminate or minimize nonvalue-added steps

A step or task that adds value to the process is one which contributes (measurably) to satisfying your customer, the user of the product or service. Therefore one of the most important steps of improving a process is to eliminate work that does not add value.

VALUE-ADDED STEP
A step that contributes to customer satisfaction. A customer would notice if it were eliminated.

It's quite common for steps and procedures to creep into processes that serve only to satisfy someone else in the organization. These are the steps that if asked why you do them, you say, "I don't know—we've always done it that way." Or, "The policy and procedure manual says we have to do it." To identify these nonvalue-added steps, look for "approval" and "for-your-information" steps in particular.

In prequality days, management's duty was to control—to make sure everyone worked according to procedure and to eliminate independent decision making. While the intent was reasonable, the effect was that a lot of work took place that had nothing to do with the customer. Bureaucratic make-work flourishes under such conditions. Now's the time to question ruthlessly the reasons for such steps. An example: Employees must complete a form and gain three levels of approval to justify a $2 long-distance call.

CUSTOMER REQUIREMENTS
The needs, wants, and expectations of your customers, in their words.

The key to using this technique is in having your customers' needs, wants, and expectations clearly defined and written—in *their* words. Otherwise, you will find yourselves arguing opinions rather than facts. In Figure 7.1, we show a step from the "Setting a table" process, with the "Develop theme or color scheme" under consideration for elimination as nonvalue-added.

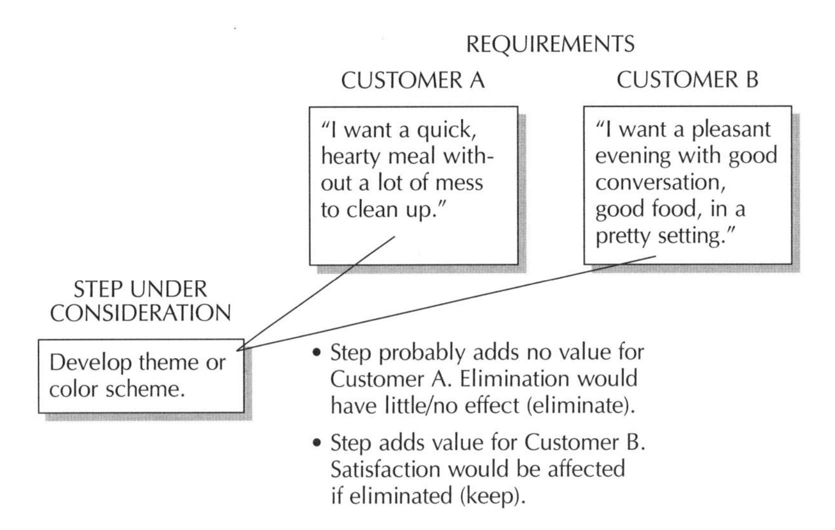

Figure 7.1. Assessing value-added.

If you didn't want to totally eliminate the step for Customer A, you might minimize or make the step less elaborate.

❶ Question the user(s) of your product or service—your customers—about their requirements. What do they need, want, and expect? What's more important to them? What's less important? Put their requirements (in their words) on a flip chart so everyone can see them, all the time.

❷ Taking one step at a time ask, "Does this step add value in our customers' eyes? What would happen to the customer if this step were eliminated?" Where you cannot justify the step, mark it. Ask others in the organization who are affected by your process if their process (and customer) would be affected if the step were eliminated. Where the answer is no, eliminate the step.

❸ A more aggressive, creative approach is to identify a particularly time-consuming step that you all agree is crucial. Brainstorm creative ways to eliminate or shorten the step—the more outrageous the ideas the better. From such discussions can come truly breakthrough ideas.

2 Develop and apply standards

Just as you developed objective questions for the decision diamonds leading to alternative paths, each inspection point must clearly specify the conditions to "pass." These objective, measurable inspection criteria are called *standards*.

PROCESS STANDARD
Precise, measurable statement of an acceptable level, quantity, or other characteristic.

In Figure 7.2, we've put an inspection point in the "Getting ready" process. We look in the mirror and make a pass/fail judgment. If we fail the mirror test, we must go back and fix—or do over—part of the process until we pass.

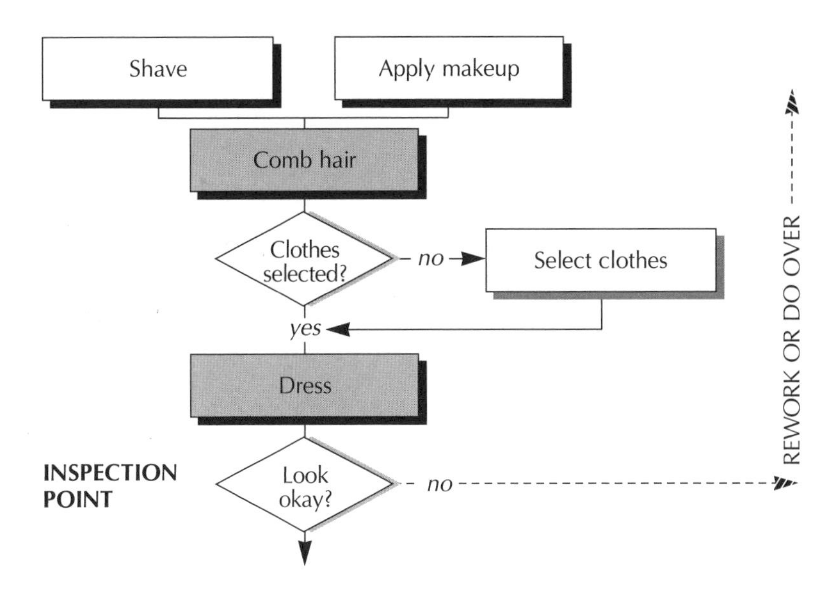

Figure 7.2. Adding inspection points.

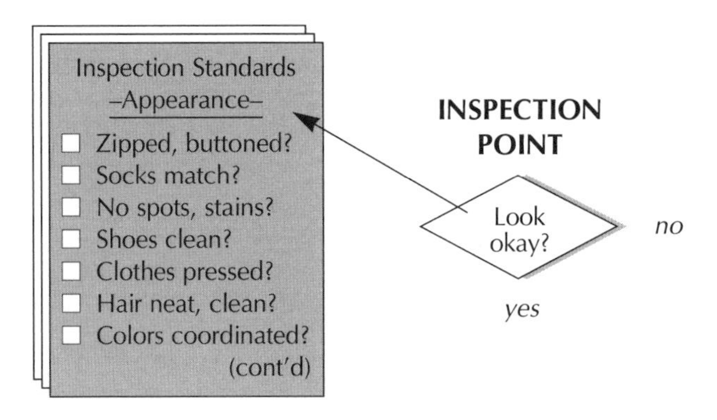

Figure 7.3. Developing standards.

However, what is the mirror test? Is everybody's mirror test the same? Do we all apply the same standards when we ask, "Look okay?" We need to operationally define what we mean by "Look okay?" Our standards might look like the ones in Figure 7.3.

The list of measurable criteria or standards does not need to be recorded on the flowchart itself, but there should be a corollary document keyed to the various inspection points. (See technique 3 for the next action in this sequence.)

❶ Select an inspection point for which you will begin developing measurable standards/criteria.

❷ Using the flip chart, brainstorm a list of possible standards.

❸ Reduce the list using the criteria of criticality, objectivity, measurability, and practicality.

❹ Agree on standards for each inspection point in the process.

3 Move inspection points forward

A key principle of quality is to prevent errors and avoid do-over and rework loops. One way to move toward this goal is to place inspection (error-catching) steps as close as possible to the point where the error occurs.

Notice the inspection standards described previously. They are applied near the end of the process—*after* shaving, combing hair, selecting clothes, and so on. If we fail the mirror test for "Hair neat, clean?" we might have to undress and go all the way back to "Wash hair." If we find a stain on our shirt collar, we'll have to undress and go back to "Select clothes," find a clean shirt, and redress.

A better way is to apply the standard (inspect) at the point where the error is likely to occur, at "Select clothes" (see Figure 7.4).

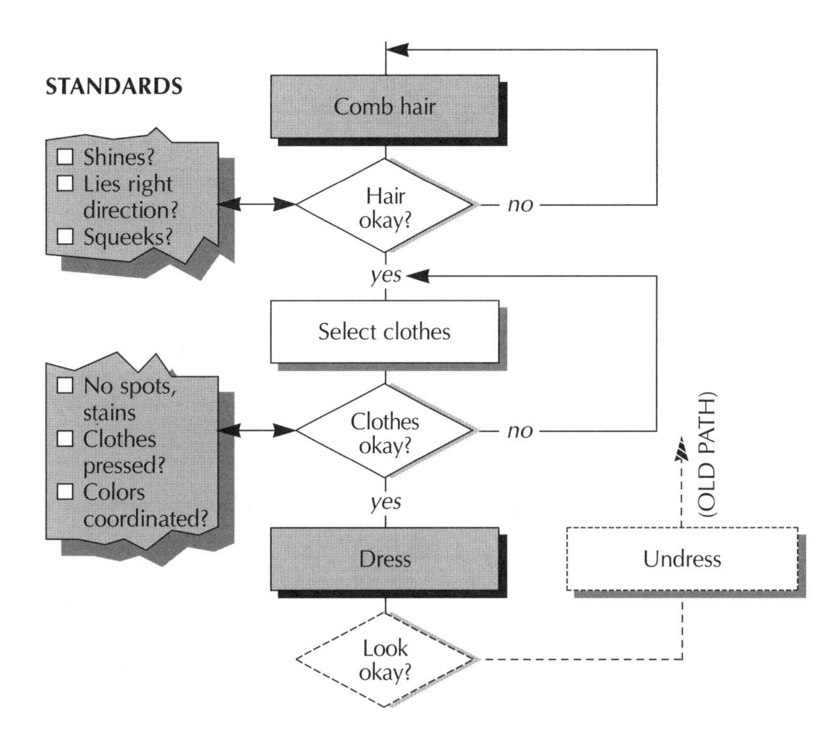

Figure 7.4. Inspection after source of error.

If we find a stained shirt, we'll replace it immediately rather than dressing, then looking for stains. The same procedure applies to our hair. We'll make sure it passes muster *before* going on to dress.

Unfortunately, we're still relying on inspection. But at least we know where the errors are occurring—an important prerequisite to eliminating errors and rework completely.

❶ Using the lists of observable standards you developed in the previous section, decide where in the process the error is likely to occur.

❷ Create an inspection point as close to the error-producing step as possible. Pair the standard with its inspection point.

❸ Decide on a practical method for catching errors in this earlier location—even if it means you must add a step to the process. Be sure your plan can (and will) be carried out by process participants. (No lip service, please.)

❹ Repeat these steps for each standard you've created.

❺ Rearrange and redraw affected sections of the map.

4

Eliminate the need for inspection points altogether

Now that you know where in the process errors tend to occur, think of ways to eliminate the possibility of error. Preventing errors takes some real creative thinking. Your first ideas may not be especially practical or effective. But stick with it. There's real payoff in this technique.

An example: How could we eliminate having to check our clothes for soil before dressing in the morning? (See the previous example.) Here are some ideas resulting from a brainstorming session:

- Buy new clothing every day. (Brainstorming is supposed to be outrageous!)
- Buy only colors and patterns that won't show soil.
- Coat all clothing with silicon spray so stains won't stick.
- Wear lab coats (or aprons) over clothing to protect them from stains.
- Buy cheap, disposable clothing. Throw away after each wearing.
- Analyze the spots over time to discover why you soil your clothing. Eliminate the cause (ink: change writing implement; food stain: use napkin or learn to eat more neatly; collar soil: wash your neck more often; and so on).
- Inspect all clothing as you take it off (moving the inspection point to another process). Soiled clothes go into a "to clean" stack and not hung up (removed from the getting ready process).
- Never wear any piece of clothing that hasn't been washed or cleaned. Solution: When taking clothes off that have been worn, put them in a hamper for dry cleaning or laundering (see Figure 7.5).

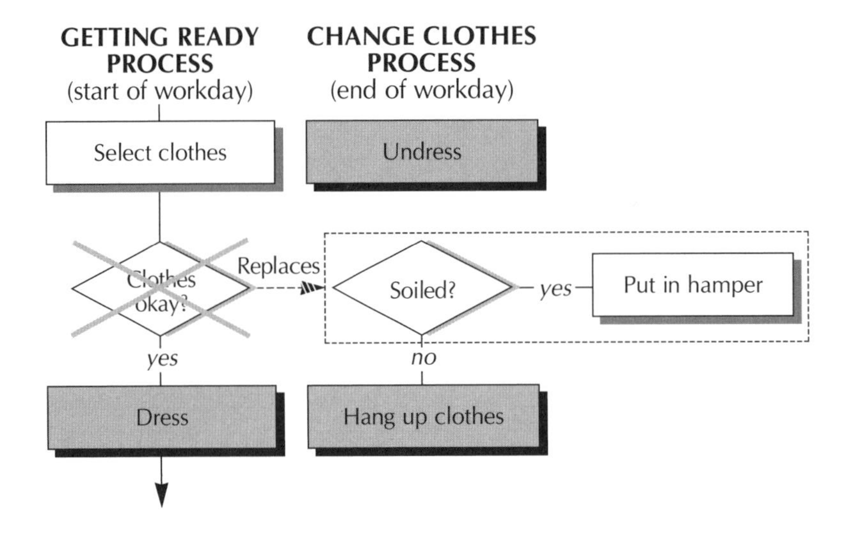

Figure 7.5. Error-prevention strategy.

Many of these options sound fairly silly, but there are several beginnings of solutions that might prove practical with some creative effort.

❶ Identify the inspection point to be eliminated.

❷ Brainstorm (without judging) a number of ideas for elimination.

❸ Evaluate each idea for its strengths and weaknesses.

❹ Develop a solution and plan its implementation.

❺ Try out the solution. Continue to monitor for errors (check) to see if your solution has worked.

❻ Rearrange and redraw affected sections of the map. Map a segment of a new process if required.

5

Chart and evaluate inputs and suppliers

INPUT
The materials, equipment, information, people, money, or environmental conditions that are required to carry out the process.

SUPPLIER
The people (functions or organizations) who supply the process with its necessary inputs.

The principle "garbage in, garbage out" is at the heart of this important technique for improvement. The quality of inputs to your process has a substantial impact on its own ultimate quality.

Whether your suppliers are internal or external to the organization, you are their customer and you should expect quality of them—just as your customers expect quality outputs from you. To be a good customer to your suppliers, you must communicate to them what you want, need, and expect (your requirements).

Figure 7.6 shows one process step, "Wash dishes" with its inputs charted. It shows that you need soap, a sponge, and hot water. Other inputs might be gloves, dishpan, dish drainer, and so on.

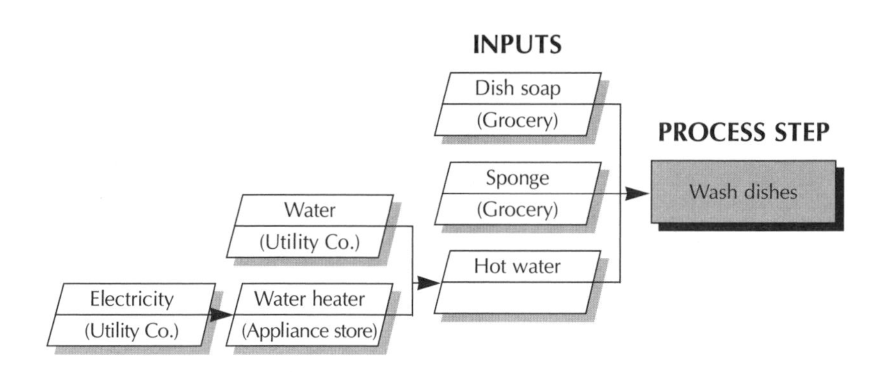

Figure 7.6. Suppliers and their inputs.

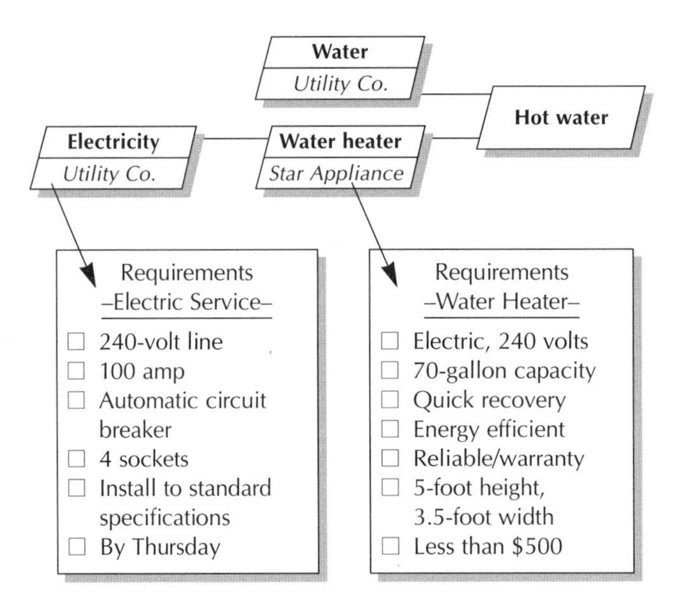

Figure 7.7. Requirements (standards) for inputs.

Inputs are written into a split parallelogram, with the input above and its supplier below. If you're using stick-on notes, use a different color to represent the parallelogram.

Because your map probably now takes up enough space for a small baseball field, you can create a separate document like the one in Figure 7.7, composed of your suppliers, inputs, and your requirements of them.

You probably won't develop all your requirements at once. As problems arise (such as not enough hot water to do a lot of dishes), you'll look to the inputs, develop your requirements, and discuss them with your suppliers. But, over the long run, paying attention to all your inputs can measurably increase both the efficiency and effectiveness of your work processes.

❶ Select process steps for which there are multiple inputs.

❷ Title a flip chart page with the name of one of the activities. Divide the page into two vertical columns. Label one "Inputs" and the other "Supplier."

❸ Brainstorm inputs. Leave suppliers blank for now.

❹ Process the list for wording, accuracy, and appropriateness.

❺ Fill in the name of the supplier for each input.

❻ Rate each input as a 1) needs attention now; 2) postpone for later action; 3) no attention needed at this time.

❼ For inputs rated "1," develop your requirements.

❽ Judge your inputs against your requirements; share findings with your suppliers. Negotiate improved inputs.

■

8

More Ways to Improve the Process

The techniques described in this chapter may be considered optional. We recommend that you consider all of them. Where you decide one is appropriate for your process, you may need a more detailed set of instructions than we are able to give here. There are plenty of good books about several of them.

Techniques in this chapter

6. Do a cycle-time study
7. Move steps into another process
8. Design a parallel process
9. Automate or mechanize step(s) in process
10. Map subprocesses
11. Use a map to train or retrain process participants
12. Get feedback on map; learn from customers, suppliers, managers, stakeholders, and other process participants
13. Use the map as a benchmarking tool

6 Do a cycle-time study

Cycle time is the elapsed time to complete the process, from boundary to boundary. Figure 8.1 shows one of several trips to the service station to get gas. We listed the major steps of the process to the left (shortened for this example) and timed each step from start to finish. The total cycle time is 19 minutes—from 9:32 to 9:51—boundary to boundary.

TOTAL CYCLE TIME
The time it takes to complete a process, from boundary to boundary. Sometimes called actual cycle time.

CYCLE-TIME STUDY		1		2	
	STATION	City Arco		Exx.	
	DAY TIME	Sat. 12/2 am		Sat/am	
		Start	End	Start	
PROCESS STEPS	Enter station	9:32		9:02	
	Drive to open pump	9:34	9:34	9:05	
	Turn off motor	9:34	9:34	9:05	
	Pump gasoline	9:36	9:39	9:08	
	Check oil	9:40	9:43	9:12	
	Pay: cash				
	Pay: credit	9:47	9:50	9:20	
	Leave station		9:51		
CYCLE TIME	Total		19		21
	Theoretical		11		12
	Diff:		8		9

Figure 8.1. Data collection sheet for cycle-time study.

However if we add the times for each step together (it took 3 minutes to pump gas, 3 minutes to pay, and so on), we have a measure of the theoretical cycle time—the amount of time actually spent to service the car (only the work), minus between-step waiting time and other bottlenecks in the process. The theoretical cycle time amounts to only 11 minutes. The difference between total and theoretical cycle time is 8 minutes—8 minutes (42 percent) of nonvalue-added time to service our car.

THEORETICAL CYCLE TIME

The sum of the times required to perform each step in the process. Does not account for hand-off or wait times. Theoretically, the shortest possible time to complete the process. The difference between total and theoretical cycle times represents the opportunity for improvement.

We waited 2 minutes for an open pump and 4 minutes (from 9:43 to 9:47) before the attendant began to process our credit card payment. In organizations that process paper, the difference between total cycle time and theoretical can amount to weeks while a paper sits on someone's desk waiting for a 10-second signature. Or a half-finished product waits for an input to arrive.

Not only is the nonvalue-added time important to minimize, with enough observations, we may find that the payment step, under the right conditions, can be done in a minute or less, but it usually takes twice that amount of time. How could we make the payment step more efficient?

To improve the cycle time of this simple example, we could decide to 1) service our car at a less busy time and 2) pay cash. These two changes could cut the cycle time from 19 minutes to about 14 minutes—considerably closer to the theoretical time. Other changes are also possible.

In general, the cycle-time study gives you the information needed to identify inefficiencies and bring the total cycle time closer to the theoretical time.

(Note: Batching work almost always contributes to cycle time. An example: Your monthly bills arrive randomly; you put them in a drawer and write all the checks at the first of the month. Some of the bills will sit in the drawer for four weeks, waiting to be paid. If you pay bills weekly, you'll cut the cycle time to 6 days, maximum. Write checks daily and you'll cut cycle time to less than 24 hours.)

❶ Develop a check sheet similar to the one in Figure 8.1. Decide how to measure (time) the steps in your process.

❷ Measure the process at least a half-dozen times, keeping other variables constant (same service station, same day of the week, same driver) to reduce the number of possible causes of the variation in time.

❸ Calculate total and theoretical cycle times for each observation (set of measurements). Calculate the difference.

❹ Look for bottlenecks and other inefficiencies that are contributing to the total cycle time. Develop solutions and try them out, taking cycle-time measures of your changed process. Adjust and adapt until your changes reliably reduce total cycle time.

7 Move steps into another process

As a means of uncluttering a process and minimizing cycle time, it's possible to move one or more steps to an earlier time, making them part of a different, less time-sensitive process. An example: Airlines used to require all passengers to pick up a boarding pass at the airport before boarding, causing long lines and grumbling passengers. Moving the step "Issue boarding pass" to the travel agent's process "Sell tickets" reduced passenger waiting time and improved customer satisfaction.

Another example: Suppose we wanted to reduce our cycle time for getting ready for work in the morning. We could select several steps from the morning process and move them to the night before (see Figure 8.2).

Now you've reduced the cycle time of the morning process by anywhere from 5 minutes to 75 minutes.

Any kind of prepreparation serves to shorten the main process; this is the principle behind the hugely profitable prepared frozen food industry. Meal preparation processes can be shortened because the food is assembled and cooked beforehand, requiring only heating during the meal preparation process.

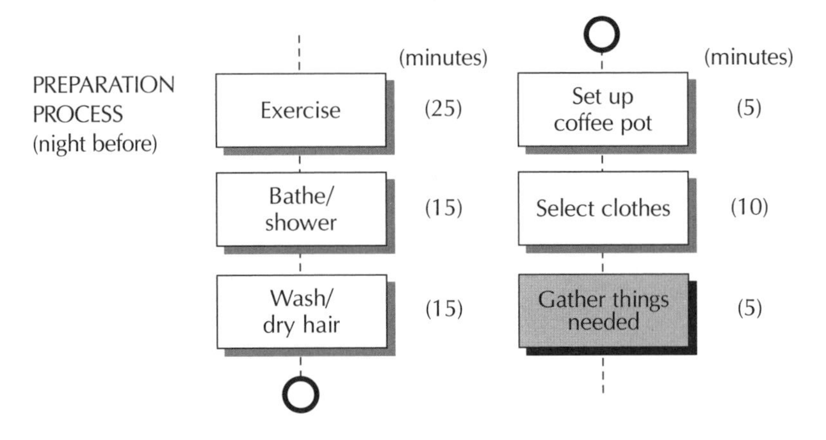

Figure 8.2. A new preparation process.

In your own process, look for anything that can be done ahead of time (without sacrificing other characteristics that are important to your customers).

8

Design a parallel process

A parallel process is one that occurs simultaneously with the primary process and usually reduces cycle time. At the grocery checkout, for example, the checker enters the price of each item; then he or she bags the groceries (see Figure 8.3).

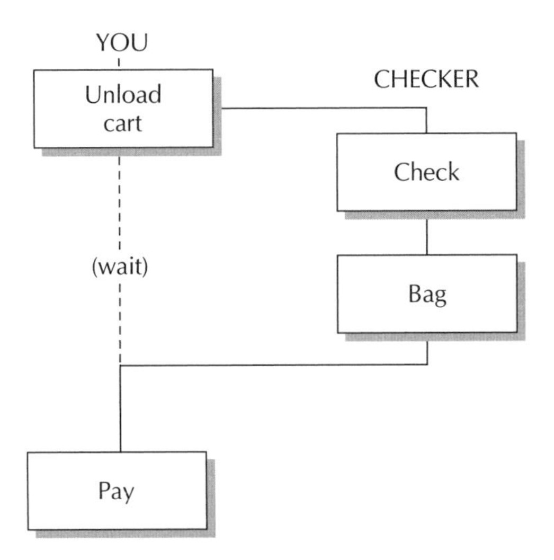

Figure 8.3. Nonparallel process, long customer wait.

But it's much faster (for you) if there's a bagger working in parallel with the checker (see Figure 8.4).

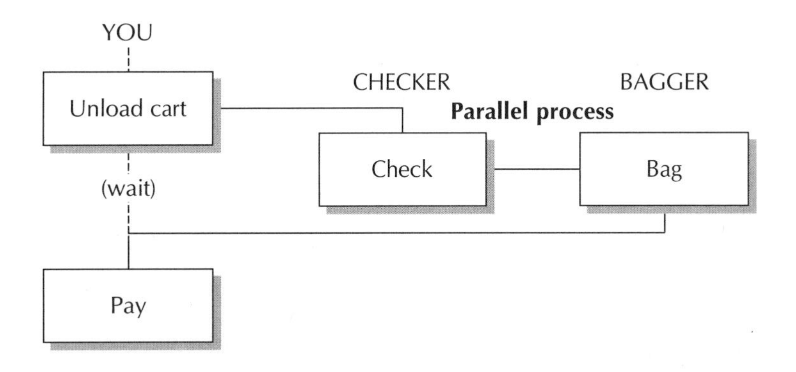

Figure 8.4. Parallel process, shorter customer wait.

Similarly, if you had the luxury of a maid, valet, or willing spouse, you could shorten your getting ready process by handing off certain tasks that could be done in parallel (see Figure 8.5).

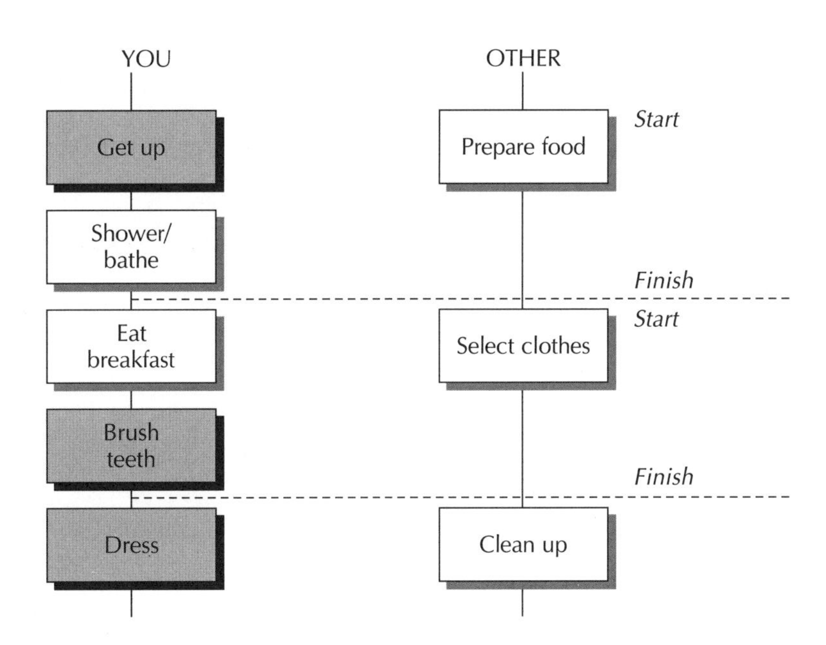

Figure 8.5. Parallel process, by person.

While you're showering, the maid/valet is preparing your breakfast. When you sit down to eat, she or he begins laying out your clothing; as you begin to dress, she or he returns to the kitchen to clean up.

As the maid/valet example demonstrates, parallel processes are time-savers, but usually require some additional resource—either a person or a machine. Consider a machine taking over a parallel process (see Figure 8.6).

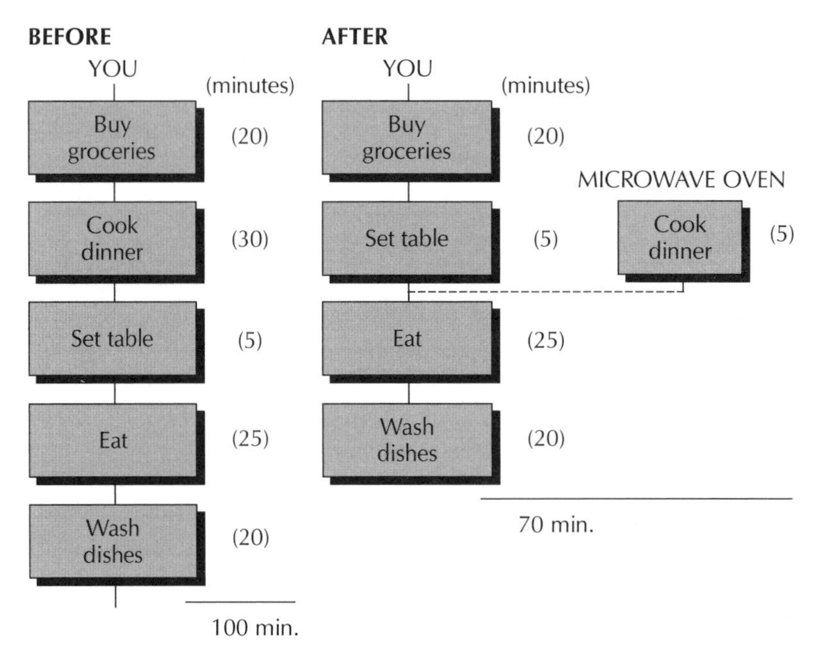

Figure 8.6. Parallel process, by machine.

By using a machine to take over part of the process, you save 25 minutes of preparation time. Watch for trade-offs, however. In spite of saving 25 minutes, the meal may not be as good.

❶ Identify steps that could be done by someone (or something) else, in parallel.

❷ Map the parallel process so that all can see exactly what would occur.

❸ Evaluate the idea by thinking of all the pluses and minuses of such a reorganization. Some criteria to consider:

- Effect on quality of the output (customer satisfaction)
- Cost
- Feasibility (including reactions of the organization)
- Unintended consequences (ripple effect on other processes)

9 Automate or mechanize steps

Like the parallel process, automation is a common method for reducing cycle time, reducing errors, or both. Computers, conveyor belts, copiers, postage machines, and bar-code readers are just a few of the machines that have taken over the multistep processes people used to perform.

Consider the example in Figure 8.7.

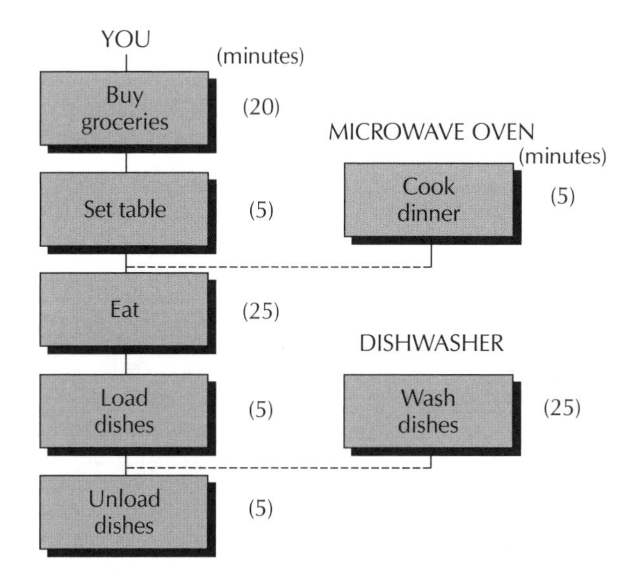

Figure 8.7. Partially automated process.

The microwave oven shortens the cycle time by 20 or more minutes. Although the automatic dishwasher actually lengthens the cycle time, it frees up the human to do other, more useful tasks (or simply to reduce the labor involved in the dinner process).

It's easy to leap to a solution that includes automation and suffer from the law of unintended consequences. Therefore it's important to analyze the effect—both long and short term—that such a solution will have. Consider the solution carefully.

❶ Identify steps that could be done by someone (or something) else.

❷ Map the new process so that all can see exactly what would occur—particularly what the eliminated human would be doing instead of processing by hand.

❸ Evaluate the idea by thinking of all the pluses and minuses of such a reorganization. Some criteria to consider:

- Effect on quality of the output (customer satisfaction)
- Cost (both long and short term)
- Feasibility (including reactions of organization)
- Unintended consequences (ripple effect on other processes)

10
Map subprocesses

The primary process with its decision points, alternative paths, and loops is likely to be rather general. It shows the major, critical steps on a macro level. Most, if not all, of the steps of the primary process can be broken down further—into subprocesses.

SUBPROCESS
The smaller steps that comprise one process step; the next level of detail. Has all the same characteristics of a primary process, such as decision diamonds, parallel processes, and inspection points.

In Figure 8.8, we see the "Do dishes" step has a number of substeps or subtasks—washing dishes is made up of scraping, stacking, washing, rinsing, and so on.

Each step in the subprocess can be further broken down into its elements and represented as a sub-subprocess. It would even be possible to take a sub-subprocess and break it into its sub-sub-subprocess.

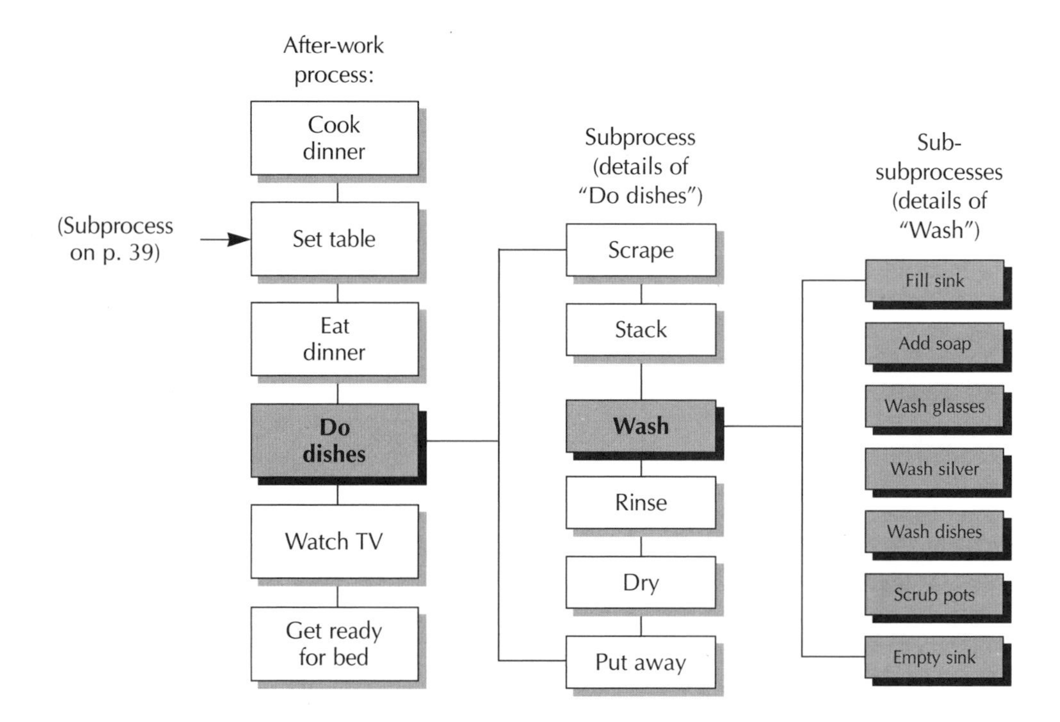

Figure 8.8. Process, subprocess, sub-subprocess.

Why would anybody ever go to that level of detail? Do you need to go to that level of detail? The answer is no—at first.

But as you progress over the months and years in your continuous quality improvement efforts, you may decide that some step of your primary process could stand some improvement. Either you've decided you're chipping too many dishes (reduce errors) or it's taking you longer than you would like (reduce cycle time).

The only way to tackle this improvement goal is to look at the "Do dishes" step in detail—by breaking out its subprocesses and perhaps even the sub-subprocesses.

Flowchart the subprocesses and sub-subs when:

- You've identified a primary step as a potential problem area
- You can find no opportunity for further improvement at the primary level
- You realize that no one understands how a particular step is actually performed—you want to understand your process even better

Although we didn't show it in Figure 8.8, subprocesses may have alternative paths, revision loops, and do-over loops, just as the primary process does.

Which leads us to the final point about primaries, subs, sub-subs, and sub-sub-subs. *The primary process is whatever you choose it to be.* In other words, on one day the primary process might be "cleaning the kitchen." Another day, you might identify the primary process as "cleaning the house" of which "cleaning the kitchen" becomes its subprocess. Yet another day, the primary process might be "household maintenance" of which "cleaning the house" is its subprocess, "cleaning the kitchen is a sub-subprocess, "washing dishes" is a sub-sub-subprocess—and so on.

❶ As a group, decide if you want/need to map any of the subprocesses, based on the previous guidelines.

If yes,

❷ Turn back to chapter 4, page 12. Beginning with "Brainstorm," repeat all the activities for the identified step.

11

Use a map to train or retrain process participants

A process map can be an ideal job aid. Either training new employees or retraining experienced employees demands that you be able to describe, step by step, what needs to be done and to what level of accuracy.

You can teach people to read a process map in about 15 minutes (it takes longer to learn how to create one). Show them the symbols and conventions and, if possible, have them practice, using the process map as a guide.

Be prepared for plenty of questions about how to make the decisions posed in the decision diamonds. Remember that it's far better to pose objective questions rather than subjective questions, as shown in Figure 8.9.

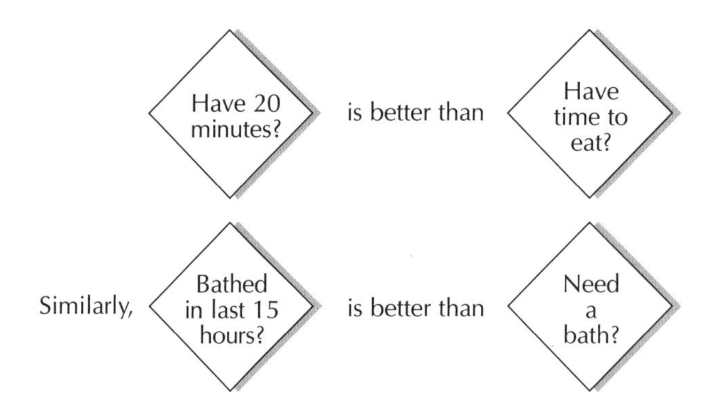

Figure 8.9. Objectivity versus subjectivity.

If your diamonds contain ambiguities, your learners will let you know—fast and vocally! Use their comments to further improve the map and its process.

Where you need to explain to process participants the changes and improvements you've made and want them to do, the map is a fine way to structure the discussion. Back on the job, you can convert the map into a checklist to help participants remember complex sequences and decisions. Just as pilots use a preflight checklist to make certain no important steps are omitted, you can use your map in the same manner.

An even more important point: Don't create a rigid, inflexible bureaucracy surrounding your process. Sure, you've done a lot more thinking and analysis about the process than anyone else. You've bled and sweated to improve it, bit by bit. But if you carve it in stone, it will belong only to you. Your objective is to make the process belong to all who work with it and to encourage *everyone* to continuously improve it.

12

Get feedback on map; learn from customers, suppliers, stakeholders, and other process participants

If you throw a ball against a wall, it will bounce back to you differently every time because the variables are almost infinite. There are a similar number of ways to look at your process. If you explain your map to dozens of people, they'll have different responses and questions every time because of the variable points of view.

Maps are a different way of looking at and talking about work sequences. People who have some kind of stake in your process will see new things from their different vantage points—things that perhaps you hadn't considered.

The objective is to collaboratively gain an even deeper understanding of how the process works—to gain new insights and to understand how your process touches and affects other processes, both inside and outside the organization. A map is a tool for the *organization* to use. Dozens of maps form a detailed system. This is true systems thinking.

Explain your map to anyone who will listen. Solicit their comments and questions. Focus on why they make particular comments or ask the questions they do.

13
Use the map as a benchmarking tool

Associated with the previous feedback technique, benchmarking is the comparison of best practices and techniques with those in organizations known to be best in class.

To avoid making benchmarking simply an exercise in organizational tourism, use the process map to guide discussions with your benchmark partner. To prepare you should:

❶ Mark important decision points and several alternative paths that you want to focus your discussion on.

❷ Ask your benchmark partner if he or she has had experience reading maps. If not, prepare a short description of how to read the map.

❸ If your map is particularly long and detailed, prepare a less detailed version, showing just the steps of the primary process and the decision points and alternative paths you highlighted in step 1.

At the meeting with your benchmark partner

❹ Overview the map broadly—an introduction, no detail.

❺ Compare value-added steps in the primary process. Does your benchmark partner do all of the steps? Are there others your partner does that you don't?

❻ One by one, introduce key decision points and their alternative paths. Be alert for differences and good ideas that you can use to change and improve your own process.

A final word

Keep good records of which techniques you try and their results. Never stop asking yourself, "I wonder why . . . ?"

Continue learning. Resolve to learn (more) about statistics. This one area of study will provide the next quantum leap in your process improvement efforts. Statistics is mostly about thinking logically and demanding proof. You don't need a degree in mathematics to understand the principles.

■

Glossary of Terms

Alternative path A path through a flowchart comprised of one or more optional tasks off the mandatory primary path; preceded by a decision diamond.

Block diagram An alternative format to a linear flowchart. Has multiple columns to account for the tasks of several people or departments. When completed, it shows how the output is passed back and forth among functions.

Boundary, process See *Process boundaries.*

Characteristics Attributes that are unique to a particular product or service. Human beings have the following characteristics: height, weight, girth, number of teeth, number of appendages, hair and eye color, foot size, gender, political affiliation, age, and so on. Generally speaking, we look for characteristics of products and services that are measurable.

Consensus Agreement, harmony, compromise. A group decision that all members agree to support, even though it may not totally reflect individual preferences. Consensus is possible when diverse points of view have been heard and examined thoroughly and openly.

Customer The person or persons who use your output—the next in line to receive it. Whether your customers are internal or external to your organization, they use your output as an input to their work process(es).

Cycle time, total The total amount of time required to complete the process, from boundary to boundary; one measure of productivity. The difference between total and theoretical cycle times represents the opportunity for improvement. (See *Theoretical cycle time.*)

Decision diamond A diamond-shaped figure in a flowchart that poses a question and signals either an alternative path or an inspection point.

Do-over loop A result of a failed inspection point, a do-over loop leads to an earlier step in the process. Steps must be repeated. Associated with scrap. (See *Rework loop.*)

Input The materials, equipment, information, people, money, or environmental conditions that are needed to carry out the process.

Inspection point A pass/fail decision, based on objective standards, to test an output in process. Signaled by a decision diamond with two or more paths leading from it. May lead to a rework loop (step) or to a do-over loop.

Macro process Broad, far-ranging process that often crosses functional boundaries (for example, the communications process or the accounting process). Several to many members of the organization are required to accomplish the process.

Mapping The activity of creating a detailed flowchart of a work process showing its inputs, tasks, and activities, in sequence.

Micro process A narrow process made up of detailed steps and activities. Could be accomplished by a single person.

Output The tangible product or intangible service that is created by the process; that which is handed off to the customer.

Parallel process A process executed by someone (or something) else that occurs simultaneously (concurrently) with the primary process. May or may not be part of the primary process.

Primary process The basic steps or activities that will produce the output—the essentials, without the "nice-to-haves." Everyone does these steps—no argument.

Process A sequence of steps, tasks, or activities that converts inputs from suppliers to an output. A work process adds value to the inputs by changing them or using them to produce something new.

Process map A graphic representation of a process, showing the sequence of tasks; uses a modified version of standard flowcharting symbols.

Process boundaries The first and last steps of the process. Ask yourself, "What's the first thing I/we do to start this process? What's the last step?" The last step may be delivery of the output to the customer.

Process owner The person who is responsible for the process and its output. The owner is the key decision maker and can allot organization resources to the process participants. He or she speaks for the process in the organization. That is, if someone says, "How come those California people aren't selling enough equipment?" the process owner—probably a District Sales Manager on the West Coast—would have to come forward to answer.

Process participants The people who actually do the steps of the process—as opposed to someone who is responsible for the process, such as the process owner/manager. For example, if you use subcontractors to produce the product, and you don't do the work yourself, the subcontractor is the process participant.

Quality Conformance to customer needs, wants, and expectations—whether expressed or unexpressed. Fitness for use.

Requirements What your customer needs, wants, and expects of your output. Customers generally express requirements around the characteristics of timeliness, quantity, fitness for use, ease of use, and perceptions of value.

Rework loop The result of a failed inspection point. A rework loop adds steps to the process and generally leads back to the inspection diamond. (See *Do-over loop.*)

Stakeholder A process stakeholder is someone who is not a supplier, customer, or process owner, but who has an interest in the process and stands to gain or lose based on the results of the process. Most processes have a number of stakeholders—such as senior managers from other departments or even government agencies.

Subprocess The smaller steps that comprise one process step; the next level of detail. Has all the same characteristics of a primary process, such as decision diamonds, parallel processes, or inspection points.

Standard Precise, measurable statement of an acceptable level, quantity, or other characteristic.

Supplier The people (functions or organizations) who supply the process with its necessary inputs.

Theoretical cycle time The sum of the times required to perform each step in the process. Does not account for hand-off or wait times. Theoretically, it is the shortest possible time to complete the process. The difference between total and theoretical cycle times represents the opportunity for improvement. (See *Total cycle time.*)

Value-added step A step that contributes to customer satisfaction. A customer would notice if it were eliminated.

Index